Audio Production Basics with Reason Software

To access online media visit:
www.halleonard.com/mylibrary

Enter Code

5456-1761-8751-0127

Audio Production Basics
with Reason Software

Zac Changnon

with contributions by

Frank D. Cook and Eric Kuehnl

NextPoint Training, Inc.

Reason 11

Rowman & Littlefield

Lanham • Boulder • New York • London

Published by Rowman & Littlefield

An imprint of The Rowman & Littlefield Publishing Group, Inc.

4501 Forbes Boulevard, Suite 200, Lanham, Maryland 20706

www.rowman.com

6 Tinworth Street, London SE11 5AL, United Kingdom

Library of Congress Cataloging-in-Publication Data available

ISBN 978-1-5381-3727-7 (paperback)
ISBN 978-1-5381-3728-4 (eBook)

♾™ The paper used in this publication meets the minimum requirements of American National Standard for Information Sciences—Permanence of Paper for Printed Library Materials, ANSI/NISO Z39.48-1992.

Contents

Acknowledgments

The author would like to give special thanks to the following individuals and groups who have provided assistance, input, information, material, and other support for this book:

Frank D. Cook, Eric Kuehnl, the band Fotograf, and the band The Pinder Brothers.

Zac would also like to thank his parents and his brother for their support, Steve Heithecker and everyone at Pyramind for their guidance and encouragement, and the crew at Hyland for their friendship.

Welcome to the World of Audio

Sounds are all around us. They make our world interesting, informative, and engaging. It's only natural to want to capture these auditory experiences—the sounds we like, the sounds we want to share, and the sounds we create. That is truly what audio production in Reason software is all about: creating, capturing, and sharing sound.

Reason Intro software provides a perfect springboard for learning the basics of audio production and improving the results of all your audio endeavors. This book has been written for readers using any edition of Reason software (Reason Intro, the full Reason software, or Reason Suite). Everything discussed in these pages applies equally to all editions of the software, unless otherwise noted. The included exercises are fully compatible with a basic Reason Intro setup. No additional hardware or software is required.

With this book, you're taking the first step toward discovering the power of Reason software and unlocking your audio creativity. Whether you've selected this book to use for self-study or have picked it up as the required text for instructor-led classroom training, you will find that it covers the principles of audio production from the ground up. It also gives you everything you need to know to fully understand the role of Reason in today's landscape of digital audio workstations (DAWs).

Getting Started in Audio Production

This book teaches the basics of recording, editing, mixing, and processing audio and MIDI using Reason software. It also provides plenty of power tips to take you beyond the basics and unleash the true power of using Reason as a creative tool. Additionally, the principles you learn in this book will apply equally to other commercial products used to create, record, edit, and process digital audio. This means you'll have a solid foundation, if you should upgrade your Reason software in the future, or switch to another digital audio workstation altogether, or even find yourself working in a full-fledged commercial studio down the line.

Who Should Read This Book

Although this book is written to support the entry-level Reason Intro software, the concepts apply equally to the full Reason edition and the expansive Reason Suite option. As such, this book serves as a resource for users of any edition of Reason software. Learners who are running full Reason or Reason Suite can use this book and complete the included exercises in a manner that is virtually identical to those running Reason Intro software.

To ensure that the topics covered in this book apply to all versions of Reason software, we have included additional details on the differences between platforms where they apply. This book is designed to help new and inexperienced users get started with little or no background knowledge. However, the scope is not limited to novice users who are experimenting with DAWs for the first time.

What Is Reason Intro?

Reason Intro is a low-cost, entry-level version of Reason software. It is available via download from reasonstudios.com for use by anyone who wishes to get started creating Reason projects on a limited budget. The Intro software is upwards-compatible with professional versions of Reason, allowing complete freedom to open Reason Intro projects on systems running full Reason or Reason Suite. However, Reason Intro is not able to open projects created with full Reason or Reason Suite.

Reason Intro runs on any compatible host computer. It does not require any additional hardware or software component add-ons. Reason Intro comes with various effects and instrument devices as well as a moderate collection of sounds and samples.

As you will learn, Reason Intro includes nearly all of the audio production features and capabilities found in full Reason software, limited primarily by a restricted track count and fewer included devices.

However, the depth of features that Reason Intro includes makes it an ideal learning environment and an excellent springboard for advanced Reason use.

About This Book

This book is designed for use in a formal course of study. The text and associated instructor-led course were developed by NextPoint Training, Inc. (NPT) as part of our Digital Media Production program and certification offerings. While this coursebook can be completed through self-study, we recommend the hands-on experience available through an instructor-led class with an NPT Certification Partner.

For more information on the classes offered through the NextPoint Training Digital Media Production program, please visit https://nxpt.us/DigitalMedia. For information on how your school can become an NPT Certification Partner, please visit https://nxpt.us/CertPartner.

Requirements and Prerequisites

This course does not require any specific background knowledge of computer systems, recording technology, or digital audio workstations. However, before starting to work with Reason software, it definitely helps to have at least a passing familiarity with computer concepts and recording gear. If you consider yourself a novice in these areas, pay special attention to the first four chapters of this book.

To try out the concepts and complete the exercises in this book, you will need to use a compatible computer and install Reason Intro software (or another edition of Reason software). Details on computer requirements are provided in Chapter 1, and installation details for Reason Intro are provided in Chapter 2.

If you will be using other connected audio or MIDI hardware, you may need to install device drivers for those components. Consult the documentation that came with your hardware or search the manufacturer's website for details.

Media Files

This book includes exercises at the end of each chapter that make use of various media files. The media files can be accessed by visiting www.halleonard.com/mylibrary and entering your access code, as printed on the opening page of this book. Instructions for downloading the media files are provided in Exercise 1.

The media files for the exercises in this book have been provided courtesy of The Pinder Brothers, Eric Kuehnl, and Fotograf.

Course Organization and Sequence

This course has been designed to teach you how to get the most out of your work with Reason software. The material is organized into 10 chapters, as follows:

- **Chapter 1. Computer Concepts**—What you need from a computer

- **Chapter 2. DAW Concepts**—What you need from your DAW

- **Chapter 3. Audio Recording Concepts**—What you need to record audio

- **Chapter 4. MIDI Recording Concepts**—What you need to record MIDI

- **Chapter 5. Reason Concepts, Part 1**—What you need to know to get started with Reason software

- **Chapter 6. Reason Concepts, Part 2**—What you need to know to work with Reason software

- **Chapter 7. Mixing Concepts**—What you need to know to mix a project

- **Chapter 8. Signal Processing**—What you can do to optimize your audio

- **Chapter 9. Finishing a Project**—What you need to do to create a stereo mixdown

- **Chapter 10. Beyond the Basics**—What to explore to become a power user

Users who have experience with computers and other DAWs may wish to skim the first four chapters, focusing mostly on the details that are specific to Reason software.

Conventions and Symbols Used in This Book

Following are some of the conventions and symbols we've used in this book. We try to use familiar conventions and symbols whose meanings are self-evident.

Keyboard Shortcuts and Modifiers

Menu choices and keyboard commands are typically capitalized and written in bold text. Hierarchy is shown using the greater than symbol (>), keystroke combinations use the plus sign (+), and mouse-click operations use hyphenated strings, where needed. Brackets ([ ]) indicate key presses on the numeric keypad.

Convention	Action
File > Save Session	Choose Save Session from the File menu.
Ctrl+N	Hold down the Ctrl key and press the N key.
Command-click (Mac)	Hold down the Command key and click the mouse button.
Right-click	Click with the right mouse button.
Press [1]	Press 1 on the numeric keypad.

Icons

The following icons are used in this book to call attention to tips, shortcuts, listening suggestions, warnings, and reference sources.

 Tips provide helpful hints and suggestions, background information, or details on related operations or concepts.

 Shortcuts provide useful keyboard, mouse, or modifier-based shortcuts that can help you work more efficiently.

 Power Tips provide shortcuts and tips for power users that can dramatically speed up your work but that go beyond the scope of the current discussion.

 Listening Suggestions refer you to audio examples that illustrate a concept or technique discussed in the text.

 Warnings caution you against conditions that may affect audio playback, impact system performance, alter data files, or interrupt hardware connections.

 Cross-References alert you to another section, book, or resource that provides additional information on the current topic.

 Online References provide links to online resources and downloads related to the current topic.

Chapter 1

Computer Concepts

...What You Need from a Computer...

In this chapter, we introduce you to the basic components of a computer system and the minimum configuration required to run Reason software. We also discuss how to configure the hardware for your system and how to configure your computer's software settings for optimal results. Lastly, we take a brief look at how to work within your digital audio workstation, using Reason Intro as an example.

⊕ Learning Targets for This Chapter

- Understand how to select a computer for your digital audio workstation

- Understand how to navigate your computer's operating system for basic file management purposes

- Understand how to set preferences for your computer and your software

- Understand how to perform basic operations with your digital audio workstation

 Key topics from this chapter are illustrated in the Reason Audio Production Basics Study Guide module available through the Elements|ED online learning platform. Sign up at ElementsED.com.

Selecting the right digital audio workstation (or DAW) involves many considerations: What features do you need? How big will your projects be? What kind of production work do you intend to do? What kind of system can you afford? And so on. Prior to making a purchase, you would be wise to check out a trial version or low-cost (feature-limited) edition of the product you are considering, if one is available.

Fortunately, for those considering Reason as their DAW of choice, Reason Studios (formerly Propellerhead) offers Reason Intro software at a reduced price. Although Reason Intro does have certain limitations, it is remarkably full featured. Reason Intro provides the same powerful workflow as the full version of Reason but with a smaller set of built-in devices and sounds and a limited number of tracks, making it an ideal learning environment for new users.

Prior to purchasing any version of Reason, you can start working with Reason for free. Reason Studios offers a 30-day trial for Reason that provides the full set of features with no limitations. In addition, even without a paid Reason license or access to the trial period, Reason software can run in Demo Mode, which allows you to test out the features of the software with certain restrictions to its use.

 See Chapter 2 for more information about using Reason software in Demo Mode.

Figure 1.1 Full version of Reason software with plug-in windows open

Before installing Reason (or other DAW software), you'll need to verify that your computer system meets the minimum requirements for the software. Once you've cleared that hurdle and installed the software on a suitable computer, you will want to configure your computer and the software application for optimal performance. This chapter will help steer you in the right direction for these and other considerations to get a system up and running successfully.

Selecting a Computer

In this section, we will look at how to select a computer for use with Reason or another digital audio workstation of your choice. We will also look at options for optimizing your computer setup to get the most out of your work environment.

Mac Versus Windows Considerations

One of your first considerations is whether to run your DAW on a Mac-based computer or a Windows-based computer. This decision is primarily one of personal preference, but it may also be based on the kind of system you already own.

If you are purchasing a new system for your DAW, understanding some general characteristics of Macs and Windows machines may help you decide on one platform versus the other.

Mac-Based Computers

One of the big selling points for Macs is how well they interact with other Apple products: iPhones, iPads, Apple TVs, and the Apple Watch. If you have already bought into the Apple ecosystem, it may make sense to purchase a Mac for that reason alone. You will find the user experience on a Mac to be similar to that of other Apple products in many respects.

Interoperability aside, Mac systems have a reputation for being easy to use and simple to understand. Additionally, Macs are known for high-quality construction and attention to detail in their design.

The downside to purchasing a Mac is the higher price you will pay for those features and conveniences. You will likely spend more for a Mac system than you would for a similarly configured Windows system.

Windows-Based Computers

Windows-based computers are available from many different manufacturers, giving you an abundance of choices when selecting a system. The competition between manufacturers means it may be easier to find a system that meets your needs at a more affordable price.

In addition to competitive pricing, Windows computers may also have an advantage in terms of the applications they support. If you use certain Windows-only applications for personal or work-related purposes, you may find that issue to be a deciding factor in your choice of platform.

Whichever platform you choose—Mac or Windows—be sure to purchase a system with adequate random access memory (RAM), processing power, storage space, and connectivity options to support your needs.

The Importance of RAM

A computer's installed RAM determines how much data and information can be stored in the computer's memory at any given time. RAM is generally used for temporary storage while an application is running. The RAM allocation is typically measured in gigabytes (or GB).

Figure 1.2 Illustration of a typical RAM module

Running Windows on a Mac

If you prefer to work on a Mac but need to run Windows for certain applications, several options are available. Modern Macs allow you to run a Windows operating system (OS) on your Mac hardware. You can configure a Mac to run Windows 10, for example, using either a separate disk partition or a virtualization software option.

To configure a separate disk partition for Windows 10, you can use the Boot Camp Assistant utility that comes with your Mac. This utility will partition your drive and guide you through the installation process for the Windows OS. If you use this option, you will need to restart your computer whenever you wish to switch between Mac OS and Windows OS.

To run Windows 10 with virtualization software instead, you can purchase and install software such as VMware Fusion or Parallels Desktop. These applications allow you to run a Windows OS on your Mac desktop without rebooting.

For either option, you will need to purchase Windows 10 separately. Factor this in when considering the total budget for your system. The virtualization software (if used) and the Windows OS may also impact the amount of RAM, storage space, and processing power required for the system.

How RAM Is Used

When you select a paragraph of text in a word processor and choose the **EDIT > COPY** command, the text is stored temporarily in RAM. You can then move your cursor to a new location in the document and choose **EDIT > PASTE**. The text stored in RAM gets added to the document at the current cursor location.

Most applications also use RAM to keep track of the work you do in your open documents. As you work, a record of your changes and edits is saved in RAM. This allows you to use the **EDIT > UNDO** command, for example, to reverse any changes you've made.

The code required to run the application is also stored in RAM while the application is running. This allows the computer to work faster than it would if it had to run the application entirely from the hard drive.

Storing data in RAM allows the information to be retrieved and operated on very quickly.

However, once you quit an application or shut down your computer, the information stored in RAM is lost. That is why you must save your changes to permanent storage first.

Why More RAM Is Better

DAWs such as Reason rely on RAM to work efficiently. Like a word processor, a DAW uses RAM for edit operations. Your DAW will also use RAM to keep track of your changes and undoable actions.

The more RAM your computer system has installed, the more memory will be available to your DAW software for temporary storage. This means the application can run more efficiently, you can copy and paste larger amounts of data, and you can maintain a longer list of complex undoable actions.

Symptoms of insufficient RAM can include sluggish behavior, frequent error messages, long pauses (or hangs) when you perform certain actions, and sudden crashes.

DAW manufacturers typically publish minimum RAM specifications required to run the application. They may also suggest that you install more RAM for more efficient operation. For work on very large or complicated projects, you may want to exceed the manufacturer recommendations.

RAM Required for Reason Intro

The minimum RAM allocation to run Reason Intro on either a Mac or a Windows computer is 4 GB, with 8 GB or more recommended. If you plan to run Reason Intro at the same time as other applications (iTunes, a web browser, an email client, etc.), you would be wise to target the high side of these numbers.

Installing More RAM

Depending on the model of computer you buy, you may be able to add more RAM at a later time, as your needs grow. This can help keep the initial purchase price down. If this is your plan, make sure to do your homework before purchasing a computer. Certain models (especially "budget" options) have RAM chips soldered in place and cannot be upgraded later with more.

Also note that upgrading RAM often requires you to replace all RAM in the computer, rather than simply adding to the existing RAM. Most computers have an even number of memory slots, with RAM assigned in pairs across them.

For example, a laptop may have 4 GB of RAM allocated across two available memory slots. In such a case, each slot will have a 2 GB memory module installed. To upgrade to 8 GB of RAM, you would need to replace each module with a new 4 GB module. (RAM modules should be added in matching pairs.)

Processing Power

The processing power, or how fast a computer runs, is a function of the computer's central processing unit, or CPU. The CPU is responsible for performing all of the processing and calculations required by an application.

Figure 1.3 The computer's central processing unit determines how fast the computer can run.

Processing Speed. The speed of a CPU is based on its clock speed. Clock speeds for modern computers are measured in gigahertz (GHz), or billions of cycles per second. A 1 GHz processor, for example, can process up to one billion instructions per second. The faster the CPU, the more powerful the computer, and the faster your applications will run.

Processor Cores. Another consideration is the number of processing cores in the computer chip. Many computer chips today offer dual cores, quad cores, or more. This means that the CPU includes multiple processing units within a single chip. The benefit of multiple cores is that the computer can process multiple instructions at one time. This in turn allows modern applications to run faster and more efficiently.

Hyper-Threading

Many CPUs support hyper-threading as an alternative (or supplement) to multiple physical cores. Intel's hyper-threading technology allows a single physical core to perform as two virtual cores. This has the same benefit as adding physical cores, as it doubles the number of instruction threads that the CPU can process at once when running an application.

How Processing Power Affects Your DAW

A host-based DAW like Reason relies on the processing power of the host computer for all audio editing, mixing, and processing. A system with insufficient processing power may exhibit sluggish performance, unwanted audio artifacts, and frequent error messages. Projects involving large track counts and heavy use of audio devices will be more susceptible to issues.

Certain audio processes are especially CPU-intensive. Some examples of common practices that may tax your CPU include the following:

- Retaining unused devices in a project or activating main mixer channel components (such as the EQ section) that are not needed

- Creating complex chains of effects devices, especially convolution reverbs and analog gear emulations

- Using many instrument devices or plug-ins in a project

- Connecting and routing devices in stereo when only a mono signal is needed

- Creating many instances of a certain effect, such as a reverb device, instead of sharing a single reverb device among multiple tracks as a send effect

To avoid running into processing limitations as your projects grow in complexity, look for a multicore CPU that runs at a reasonably high speed.

Processing Requirements for Reason Intro

Reason Studios does not specify a minimum CPU speed requirement for Reason Intro but instead simply indicates the need for a multicore Intel processor. Alternatively, a multicore AMD processor can be used with a Windows system. However, an Intel Core i7 dual-core processor running at 2.0 GHz or better (or a comparable AMD processor in a Windows system) is recommended.

Most computer systems available today meet the minimum CPU requirements to run Reason Intro.

 Reason Intro supports multithreading, so you will get better performance out of a quad-core processor than you will out of a dual-core, for example.

Always Check Compatibility Before You Buy!

New versions of Reason often have more stringent compatibility requirements than previous versions. Prior to buying a system, be sure to check the compatibility requirements on the Reason Studios website.

Requirements for Reason 11 can be found at the following web address:

- https://help.reasonstudios.com/hc/en-us/articles/360002215674-What-are-the-minimum-system-requirements-for-Reason-11-

---

Storage Space

Another important consideration when buying a computer is the amount of storage space the computer provides. This is a function of the computer's installed hard disk drive (HDD) or solid state drive (SSD). The size of the storage drive determines how many applications you can install and how much space you will have for saving files on your system.

Figure 1.4 Illustration of a hard disk drive

Determining How Much Drive Space You Need

When determining storage requirements for your DAW, you will need to consider multiple factors:

- How much space is required to install the DAW software application itself

- How much space is required to install the plug-ins you intend to use with the DAW

- How much space you will need for the projects and media files that you create with the DAW

- How much space you will need for other media files that you may want to use with the DAW (such as sample libraries, loop libraries, and sound effects)

In addition, you may need to consider the impact of other files and applications that you want to use on your computer. For example, if you plan to store your digital photos and your personal music collection on the computer, you will need to allocate additional space for those purposes.

Likewise, if you will use the computer for work in programs such as Microsoft Word, PowerPoint, and Excel, you will need additional space to install each of those applications.

Installation Requirements for Reason Intro

To install Reason Intro software, you will need a minimum of 4 GB of storage space for the application. Reason Intro may also need up to 20 GB in *scratch* storage space. This is storage space used to temporarily hold recorded audio that has not yet been saved into a song project file, as well as audio data that is generated from operations such as time stretching.

The drive used to provide scratch storage space can be changed in Reason's preferences.

 The full version of Reason requires an additional 8 GB to install included optional devices and sound library content, totaling 12 GB of required storage space. Reason Suite adds another 12 GB of optional devices and content, for a total of 24 GB required to fully install the software.

Storage Options

Many storage options are available for today's computer systems. A computer's internal system drive may be either an HDD or an SSD. The internal storage can be supplemented by any number of external drives and storage options. To allow your files to be accessed from anywhere, you can also consider cloud-based storage.

HDD Versus SSD Storage

HDD. Hard disk drives use spinning metal disks to store data. The more disks in the drive, the higher the data capacity. Storage sizes for modern HDDs typically range from around 1 to 4 terabytes (TB).

The data transfer speed for an HDD is based in part on the speed of the spinning disks, measured in revolutions per minute (RPM). Common options include 5400 RPM drives and 7200 RPM drives.

Hard disk drives provide greater storage capacity at a lower price than their solid state counterparts.

SSD. Solid state drives, by comparison, use flash memory instead of spinning metal disks for storage. The storage capacity of an SSD is based on the number and size of memory chips it has. Storage sizes for SSDs typically range from 256 GB to 1 TB.

Solid state drives provide faster access speeds than HDDs, meaning they provide better read/write performance for the computer's functions. This is noticeable in faster startup times for the computer, faster launch times for applications, and faster file transfers among systems and drives.

Solid state drives also require less power than HDDs. This gives them an advantage for laptop computers and mobile devices. Additionally, SSDs have no moving parts, so they operate silently and are less prone to mechanical failure. These advantages come at the expense of higher cost and lower overall storage capacity.

External Drive Storage

A popular option for supplementing a computer's storage is to use an additional external drive. External drives are readily available, offering a variety of connection types to match the available ports on your computer. Some common connection types include the following:

- FireWire 400 or 800
- USB 2.0 or 3.0

- USB-C

- eSATA

- Thunderbolt 1, 2, or 3

If you plan to store media files on an external drive for use with your DAW, look for a high-performance drive with a high-speed data connection to your computer. Good drive options would include a 7200 RPM HDD or an SSD. Good data connections would include USB 3.0, USB-C, eSATA, Thunderbolt 2, or Thunderbolt 3. Be sure to match the capabilities of your computer when selecting the connection type.

Cloud Storage

Both the internal system drive of your computer and any connected external drives are considered local storage. Your local storage can be complemented by any number of cloud-based storage options.

Cloud storage utilizes large-capacity storage servers and data centers that you connect to via the Internet. Cloud storage options generally require a subscription. To allow for quick access to your cloud-based content, cloud storage services usually allow you to sync your cloud content to your local storage.

The advantages of using cloud storage are many. For example, cloud storage enables you to keep your files in sync across multiple devices, to easily share files with anyone who has an Internet connection, and to retrieve your files from any computer, anywhere in the world.

Some popular cloud-based storage services include Dropbox, iCloud Drive, and Google Drive.

Onboard Sound Options (Audio In and Out)

Another aspect to consider when selecting a computer is the onboard sound options that the model offers. Many DAWs, including Reason Intro, can utilize the computer's built-in sound options for recording and playback. This allows you to use any built-in microphone inputs on the computer for recording to your project and to use the built-in computer speakers for playing back your project.

While you may not want to use the computer's microphones for any meaningful recording project, having onboard audio will allow you to run your DAW without requiring a connected audio interface. This can save you some money at the outset. It also provides the flexibility to work on the go, especially when using a laptop, without needing to bring along additional hardware.

Although audio inputs are not required, audio outputs are critical. Look for a computer that has built-in speakers and/or a stereo headphone jack for monitoring playback from your DAW.

Other Options to Consider

Aside from the computer hardware itself, you may also want to consider options for peripheral devices. With the right accessories, you can customize your setup for efficiency and/or personal preference.

Mice and Trackballs

Many computer users find the mouse that comes with their computer to be less than ideal for long-term use. Numerous alternatives are available, offering features such as multiple buttons, scroll wheels, trackballs, gesture/touch input, and wireless connection to the computer. Mouse upgrades can range from around $20 to upwards of $100. Be sure to try out any options you are considering while running your DAW (or a similar application) to determine what works best for you.

Trackpads and Touchpads

As an alternative or supplement to a traditional mouse, Apple offers several trackpad options (such as the Magic Trackpad and Magic Trackpad 2). These devices provide touch-based gesture input, similar to that found on many modern laptops. Logitech, Dell, and others offer similar touchpad devices for Windows computers.

Extended Keyboards and Keypad Options

An upgrade option that can be valuable is an extended keyboard. The extended keyboard provides access to numeric keypad keys on the righthand side, which adds an easy way to access several commands in Reason.

Figure 1.5 Standard Mac keyboard (top) and extended Mac keyboard (bottom)

Benefits of an Audio Interface

Many audio interfaces are available that are compatible with Reason Intro. Any audio interface that is Core Audio–compliant (Mac) or ASIO-compliant (Windows) will work with Reason Intro.

Some good options include the **Focusrite Scarlett** series or **Focusrite iTrack Solo**. Many Focusrite interfaces are also available in **Studio** editions, with microphone and headphones included.

The benefits of using an audio interface can include better sound quality, a greater number of available inputs and outputs (I/O), and access to digital I/O options. For many beginning users, the number of inputs may be the most significant issue.

Reason supports a maximum of 64 audio input channels with compatible hardware, although Reason Intro supports only 16 total tracks. Any number of available audio tracks can be record-enabled simultaneously, but the processor and storage performance of your computer will limit the number of tracks you can simultaneously record in practice. Without an audio interface, you are limited to the input options available on your computer.

Working with Your Computer

Once you have selected an appropriate computer to host the DAW of your choice, you should spend some time to get familiar with basic operations on the system. If you already have experience with computers and the operating system you will be using, you may want to skip or skim this section. The information in this section focuses on navigating among the files, folders, and applications on your computer.

In this section, we cover basic operations such as locating, moving, organizing, and saving files, creating folders, and launching applications.

File Management

Perhaps the most important thing you should know about your computer is how to navigate among the files and folders using the computer's operating system. On the Mac, you will use the Mac Finder for this purpose. On a Windows computer, you will use File Explorer windows.

Using the Finder (Mac-Based OS)

The desktop experience in a Mac-based operating system is defined by the Finder application. The Finder is the first thing you see after you start up a Mac, before you launch any applications or open any files or folders.

Figure 1.6 The Finder icon on a Mac

The Finder includes a menu bar at the top of the screen, the desktop display in the center (a background image along with any desktop icons and open Finder windows), and the Dock at the bottom of the screen.

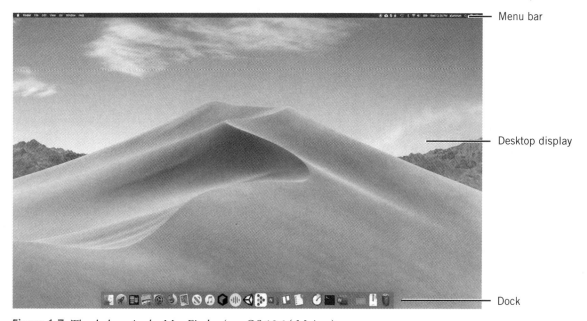

— Menu bar

— Desktop display

— Dock

Figure 1.7 The desktop in the Mac Finder (macOS 10.14 Mojave)

The Finder Versus the Desktop

Although the terms are sometimes used interchangeably, the Finder and the desktop are two different things. The Finder is an application on the Mac that is always running. You can switch to the Finder application at any time, the same way you switch between other running applications.

The desktop is the default location displayed by the Finder. You can think of the desktop as the background canvas upon which your open folders and applications are displayed. The desktop also provides a location to store folders and files.

The items stored on the desktop are visible from the Finder. They can also be displayed in a Finder window, by opening the Desktop folder.

From the Finder, you can:

- Navigate through the file system on your storage drive(s)

- Open folder locations

- Manage files within open folders

To open a new Finder window, choose **FILE > NEW FINDER WINDOW** from the menu bar at the top of the screen (or double-click on a folder on the desktop). You can navigate within an open Finder window by clicking on a location in the sidebar or by double-clicking on a displayed folder in the file list to open it.

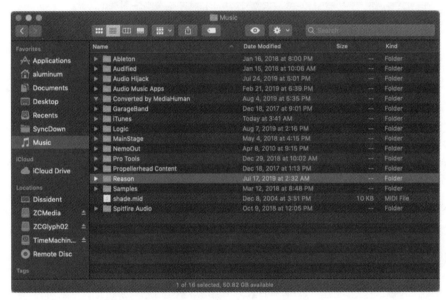

Figure 1.8 An open Finder window (macOS)

Using File Explorer (Windows-Based OS)

Windows computers also provide a desktop view (similar to the Mac Finder). However, the desktop might not be the first thing you see upon startup, depending on the operating system you are using. (Windows 8 and Windows 8.1 use the Metro UI tile view by default, similar to that on a Windows Phone.)

To open a new File Explorer window, do one of the following:

- In Windows 7 or Windows 10, click on the yellow **FILE EXPLORER** icon in the taskbar at the bottom of the desktop screen (or double-click any folder on the desktop).

Figure 1.9 Mouse cursor next to the File Explorer icon in the taskbar (Windows 7 shown)

- In Windows 8 or Windows 8.1, first click the **DESKTOP** tile from the tiled Start screen; then, from the desktop, click the **FILE EXPLORER** icon in the taskbar (or double-click any folder on the desktop).

 In Windows 8.1, you can also access the File Explorer directly from the tiled Start screen by typing *file explorer* and clicking on the File Explorer icon that appears.

As in the Mac Finder, you can navigate within an open File Explorer window by clicking on a location in the Navigation pane on the left or by double-clicking on a displayed folder in the file list to open it.

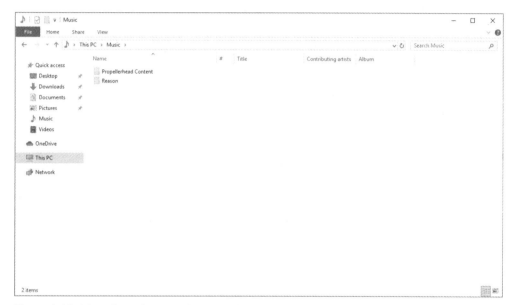

Figure 1.10 The File Explorer window (Windows 10)

Locating, Moving, and Opening Files

Using the locations in the sidebar of a Mac Finder window (or the Navigation pane in File Explorer), you can navigate to different folder locations on your computer. For example, clicking on the **DOWNLOADS** location will take you to the Downloads folder on your system (for any files you've downloaded from the Internet). Clicking on the **DOCUMENTS** location will take you to your Documents folder.

You can access the files within a folder at a given location by double-clicking on the folder to open it. On a Mac, you can also click on the triangle next to a folder to expand it and display its contents.

Figure 1.11   Finder window with several folders expanded

Moving Files. To move a file from one location to another, you can drag the file from its current location to any other visible folder within the window or to any location displayed in the sidebar/Navigation pane. You also can drag a file from one open Finder window or File Explorer window to another.

Opening Files. Opening a file from a Finder or File Explorer window is usually as simple as double-clicking on the file. In most cases, the file will open in the application that created it, if available on your computer. If the application is not available—or if your computer cannot determine an appropriate application to use for the file—you can choose a compatible application on your system. On both Mac- and Windows-based systems, you can right-click on a file and select an application to use from the right-click menu.

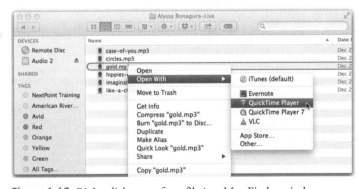

Figure 1.12   Right-click menu for a file in a Mac Finder window

Accessing Right-Click Functions on a Mac

For users who are new to the Mac, using the mouse can require an adjustment. The Apple mouse doesn't offer separate left- and right-click buttons. Instead, you must rock the mouse to get the desired effect. For a standard left-click, you must click on the *left side* of the mouse, causing the mouse to rock the left. For a right-click, you need to click on the *right side*, causing it to rock to the right.

To improve your results, practice using two fingers on the mouse. Place your index finger on the left side and your middle finger on the right. (Reverse if you are left-handed.) For a left-click, roll your hand to the left and click with your index finger. For a right-click, roll to the right and click with your middle finger.

If you are using a Mac trackpad instead of a mouse, use a two-finger click to access right-click functions.

Creating Folders

On both Windows and Mac, you can easily create a new folder on the desktop or at another desired location. While displaying the desktop or viewing a location in a Finder or File Explorer window, simply right-click with the mouse and select **NEW FOLDER** (Mac) or choose **NEW > FOLDER** (Windows) using the pop-up menu. (See Figures 1.13 and 1.14.)

You can also create a new folder by using a menu command (Mac) or Ribbon button (Windows 8 and later) or by using associated keyboard shortcuts: **COMMAND+SHIFT+N** (Mac) or **CTRL+SHIFT+N** (Windows).

Figure 1.13 Creating a folder by right-clicking on a Mac

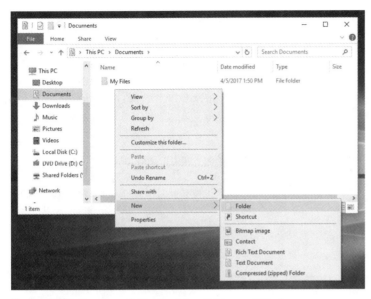

Figure 1.14 Creating a folder by right-clicking on Windows

Launching Applications

To launch an installed application, you can either double-click on the app icon in your computer's Applications directory or use one of the shortcut operations provided by your OS.

Mac-Based Systems. The easiest path to opening an application on the Mac is the Dock. The Dock is a bar of icons that sits at the bottom of your Mac screen by default. It provides quick access to your installed apps. To launch an app, simply click its icon in the Dock.

Figure 1.15 Launching Reason Intro from the Dock

Windows-Based Systems. One easy way to launch an application on any Windows system is to use the Search function. Simply press the **START** key (also known as the **WINDOWS** key) and begin typing the name of the application you wish to open. (See Figure 1.16.)

(i) Pressing the START key followed by any text activates a Start Menu search in Windows 7, a search in a right-hand pane in Windows 8, or a Cortana search in Windows 10.

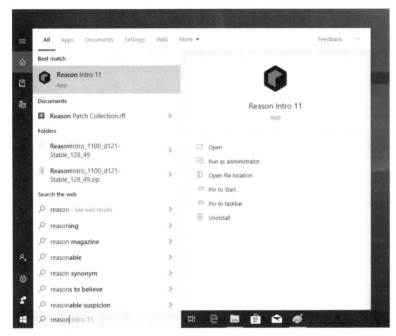

Figure 1.16 Using a Cortana search to locate Reason in Windows 10

Launching an App from the Desktop

During the install process, many applications will offer the option to place a shortcut icon, or *alias*, on your desktop. This shortcut links to the application file in your Applications directory.

Figure 1.17 Reason Intro shortcut on the Windows 10 desktop

Shortcut icons provide a convenient option for launching an application, as you can simply double-click on the shortcut from the desktop to start the application.

Working with an Application

With your application up and running, you will be able to begin working on a project—whether that be a slide presentation, a spreadsheet document, or an audio recording project. Most modern applications, including Reason, use menus and keyboard shortcuts for accessing their main commands.

Menus

The term *menu* is commonly used in software to denote one of the pull-down lists of commands at the top of the application. On Windows computers, the menu bar is typically anchored to the top of the application window. On Macs, the menu bar is typically anchored to the top of the screen.

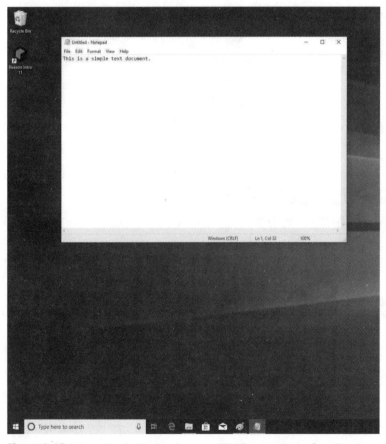

Figure 1.18 Menus in the Notepad app on Windows 10 (top of application window)

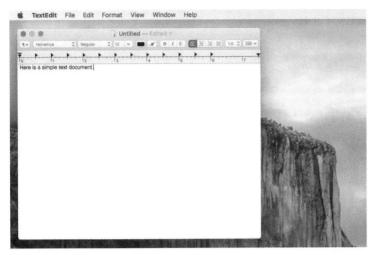

Figure 1.19 Menus for the TextEdit app on Mac OS X 10.11 (top of the screen)

To select a command, click on the associated menu name in the menu bar (such as the **FILE** menu) and then click on the desired command from the pull-down menu (such as the **SAVE** command). In some cases, menu items will include a submenu that you can use to select from multiple choices.

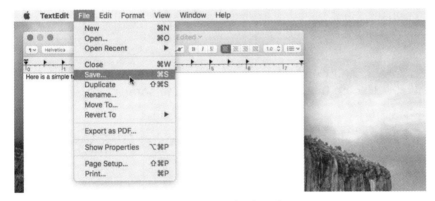

Figure 1.20 Selecting the Save command under the File menu

At times, you may have to distinguish between menus used by different applications or by the operating system (such as the Mac Finder). On a Windows-based system, you will likely find it easy to tell which menus belong to a given application, since the menus are anchored to the application window.

On a Mac-based system, however, you can run into situations where an application's windows are closed, but the application is still active. In these cases, you may see the desktop or another running application on the screen, making it less apparent which application is active.

Figure 1.21   Reason Intro menus displayed on a Mac with all application windows closed

To determine which application is active on a Mac, look for the application menu on the left side of the menu bar (immediately to the right of the Apple menu). The name of the active application will always display, regardless of whether the application has any windows open.

 Clicking on the desktop screen, an open Finder window, or a background app will make that item active, replacing the menus at the top of the screen.

 You can toggle through open applications by pressing COMMAND+TAB on a Mac or ALT+TAB on Windows.

Keyboard Shortcuts

As an alternative to using the menus to access commands, you may also be able to use keyboard shortcuts. Keyboard shortcuts use one or more modifier keys in combination with a standard key press to activate a command. Common examples include **COMMAND+S** on the Mac (**CTRL+S** on Windows) to save the open file or **COMMAND+C** on the Mac (**CTRL+C** on Windows) to copy a selection.

Mac and Windows systems each use different modifier keys, but their functions are similar. Table 1.1 shows the modifier keys available on each platform.

Table 1.1 Modifier keys available on Mac and Windows systems

Modifier Keys on Mac-Based Computers	Modifier Keys on Windows-Based Computers
The **COMMAND** key (⌘)	The **CTRL** key
The **OPTION** key (⌥)	The **ALT** key
The **CONTROL** key (^)	The **START** key (⊞)
The **SHIFT** key (⇧)	The **SHIFT** key

The advantages of using keyboard shortcuts include speed and efficiency, as well as reduced dependency on the mouse. This not only makes your work more precise, but it also helps you avoid developing ergonomic problems, including carpal tunnel syndrome.

Often, keyboard shortcuts are listed next to the command names in the pull-down menus. You'll find it worth your time to memorize the shortcuts for some of the commands you use frequently in your DAW. In addition to the shortcuts shown in menus, Reason Intro includes numerous other hidden keyboard shortcuts. We will be covering many of these in this book.

Review/Discussion Questions

1. What are some reasons for choosing a Mac computer for your DAW? What are some reasons for choosing a Windows computer instead? (See "Mac Versus Windows Considerations" beginning on page 3.)

2. How is RAM used by an application? Is RAM used for permanent storage or temporary storage? (See "How RAM Is Used" beginning on page 4.)

3. What are some indications that a computer system does not have enough installed RAM for the applications you are using? (See "Why More RAM Is Better" beginning on page 5.)

4. What factors affect how much processing power your computer has for running applications? (See "Processing Power" beginning on page 6.)

5. What are some examples of audio processes that can strain the CPU of an underpowered system? (See "How Processing Power Affects Your DAW" beginning on page 6.)

6. What are some things you should consider to determine how much storage space is required for your DAW computer system? (See "Determining How Much Drive Space You Need" beginning on page 8.)

7. What is the difference between an HDD and an SSD for storage? What are some advantages of each? (See "Storage Options" beginning on page 9.)

8. How are built-in sound options on a computer useful for Reason Intro? (See "Onboard Sound Options" beginning on page 10.)

9. How can you quickly navigate to different folder locations on your computer? (See "Locating, Moving, and Opening Files" beginning on page 15.)

10. What are some ways to create a new folder at the current location on your computer? (See "Creating Folders" beginning on page 17.)

11. How can you select a command in an application such as Reason Intro? (See "Menus" beginning on page 20.)

12. What modifier keys are available on Mac computers? What modifiers keys are available on Windows computers? (See "Keyboard Shortcuts" beginning on page 22.)

To review additional material from this chapter and prepare for certification, see the Reason Audio Production Basics Study Guide module available through the Elements|ED online learning platform at ElementsED.com.

Exploring Audio on the Computer

🎧 Activity

In this exercise, you will download the media files used for this course and move them to an appropriate storage location on your system. Then you will use your computer to listen to sample files for an excerpt from the song "Overboard" by Sacramento-area band The Pinder Brothers.

🕐 Duration

This exercise should take approximately 10 minutes to complete.

⊕ Goals/Targets

- Practice file management techniques

- Practice critical listening techniques

- Explore sonic elements that comprise a finished mix

Exercise Media

This exercise uses media files taken from the song, "Overboard," provided courtesy of the band The Pinder Brothers.

Written by: Matt Pinder; Performed by: The Pinder Brothers;
Produced by: Scott Reams and The Pinder Brothers*

The media provided for this course may be used for educational purposes only. No rights are granted to use the media for any other personal, commercial, or non-commercial purposes.

* The mix, processing, and media files have been adapted for use in the exercises contained herein.

Getting Started

To get started, you will download the media files for this course and locate them in your Downloads folder. You will then move the downloaded media files to your Documents folder or another appropriate location.

Download the course media files:

1. Launch the browser of your choice on your computer (Safari, Microsoft Edge, Firefox, Chrome, etc.).

2. Point your browser to www.halleonard.com/mylibrary and enter your access code (printed on the opening page of this book).

3. Click the **DOWNLOAD** link next to the Reason APB Media Files listing in your My Library page. The Reason APB Media Files folder will begin downloading to your Downloads folder.

4. Open the Downloads folder on your system:

    -   On a Mac, activate the Finder and choose **GO > DOWNLOADS**.

    -   On a Windows computer, press **CTRL+J** from within your browser and then click the link labeled **OPEN FOLDER**, **SHOW IN FOLDER**, or similar.

Move the Reason APB Media Files folder to a permanent location:

1. Select the Reason APB Media Files folder in the Downloads folder and drag onto the Documents location displayed in the sidebar/Navigation pane of the Finder or File Explorer window.

2. Click on the Documents location in the sidebar/Navigation pane to open the Documents folder.

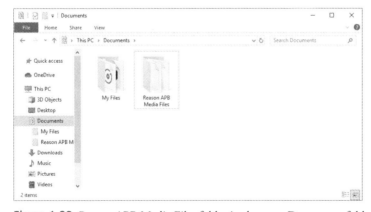

Figure 1.22 Reason APB Media Files folder in the open Documents folder (Windows 10 shown)

Listening to Sample Files

In this part of the exercise, you will locate and play various audio files. In the process, you will practice listening for different musical parts in each file.

Listen to the music mix:

1. Double-click on the **Reason APB Media Files** folder in the open Finder or File Explorer window to open the folder and display its contents.

2. Double-click on the **01. Pinder Brothers** folder to open it. You will see the following files in the folder:

    *   OverboardMix.wav
    *   Stem1.wav
    *   Stem2.wav
    *   Stem3.wav
    *   Stem4.wav

3. Select the **OverboardMix.wav** file.

4. Do one of the following to listen to the selected file:

    *   On a Mac-based system, press the spacebar to begin playback using the Quick Look feature.

    *   On a Windows-based system, right-click on the file and select **PLAY WITH WINDOWS MEDIA PLAYER**.

5. As you listen to the mix, listen for each of the different instruments and sounds that you can hear.

    Using the table below, write down each part you hear. Rate the prominence of the part in the mix on a scale from 1 to 5, with 5 being very prominent/easy to hear and 1 being barely noticeable.

Instrument or Part	Prominence
Lead Vocal	5

Listen to the stem files:

1.   Select the **Stem1.wav** file and listen to it using the same method that you used in Step 4 above.

2.   Write down the main part you hear in the stem file, along with a short description, using the table below.

3.   Listen for additional parts in the stem file and use the table to write down each different part you hear along with a short description of each.

4.   Repeat the process using each of the other stem files (Stem2, Stem3, and Stem4).

      Not all of the stems will have different discernable parts. If you are not accustomed to picking out and describing the different musical sounds you hear, just do your best to make distinctions where something sounds different. This is good ear training regardless of what type of audio production you are interested in.

      Some terms you might use in your descriptions include the following: lead vocal, background vocal, vocal harmony, vocal double, rhythm guitar, lead guitar, bass guitar, kick drum (bass drum), snare drum, crash cymbals, hi-hat, ride cymbal, and so on.

Stem File	Description of the Instrument or Part
Stem1	Lead vocal – Main vocal melody throughout

Finishing Up

After you've listened to each of the stem files, return to the **OverboardMix.wav** file and listen through it again. Are you better able to distinguish each part in the mix, now that you've heard them in isolation?

You should now be able to clearly pick out things such as the cymbal crashes and vocal harmonies that you may have missed before. You may be able to distinguish subtler details as well. Can you hear what the bass guitar is doing during the lead guitar part? Can you hear individual kick and snare drum hits? Can you tell where the hi-hat is being played and where it switches to the ride cymbal?

 To hear individual drum parts in isolation, check out the audio files in the z-Drum Parts folder. These files will help you learn what each drum piece sounds like and let you know what to listen for in a drum kit.

That completes this exercise. For more practice honing your listening skills, try identifying parts in popular songs that you hear on the radio or Spotify. If you are more interested in post-production applications, listen for different sound elements in your favorite TV program. Examples include dialogue, background ambience (birds chirping, waves crashing), Foley (footsteps, character movements, noise from objects being handled), sound effects (telephones ringing, gun shots, laser beams), and background music.

Chapter 2

DAW Concepts

...What You Need from Your Digital Audio Workstation...

This chapter introduces you to the basic concepts of the DAW. We begin by looking at some of the popular DAWs in use today. Then we explore the common audio plug-in formats that are used to add both effects and virtual instruments to a DAW. Next we look at the installation and initial setup of a Reason Intro system. We conclude this chapter with a brief explanation of some important terminology that you'll need to know to work effectively with Reason software.

✛ Learning Targets for This Chapter

- Learn the basic concepts of digital audio workstations

- Gain an understanding of common plug-in formats

- Install and run Reason Intro

- Become familiar with important audio terminology

 Key topics from this chapter are illustrated in the Reason Audio Production Basics Study Guide module available through the Elements|ED online learning platform. Sign up at ElementsED.com.

Although this book is written for Reason Intro, it is important for you as an audio production enthusiast to have at least a passing familiarity with other digital audio products on the market. Not only are you likely to encounter audio practitioners who favor these platforms, you also may end up using some of them yourself from time to time, to supplement your work in Reason.

Functions of a DAW

The term DAW stands for digital audio workstation. This term is used generically to refer to any software that can be used to record, edit, and mix audio, although most modern systems can be used for MIDI recording and editing as well.

What Can a DAW Do?

Some DAWs excel at audio editing, while some are primarily known for their MIDI features, but almost every modern DAW can do both to some degree. Reason is generally considered to be a superior platform for working with MIDI content. However, Reason also provides a robust platform for audio work.

In addition to basic recording and editing functions for audio and MIDI, DAWs often provide sound modules (or virtual instruments) for use with MIDI data. Such virtual instruments are commonly available in the form of devices and plug-ins that use MIDI data to trigger sounds.

Most DAWs also use devices and plug-ins for audio processing, to apply effects and polish to audio recordings. Additionally, DAWs typically provide mixing and automation functions for blending tracks, creating an effective stereo image, and incorporating dynamic changes during playback.

Common DAWs

Many DAWs are available, ranging in price and feature set. Here, we provide an overview of some of the most popular DAWs and highlight some of the differences between them.

Reason

Reason was first released in 2000 with the goal of providing a cost-effective, all-in-one production workstation focused on electronic music. The initial version contained a number of built-in devices, including a virtual analog subtractive synthesizer, a sampler, a drum machine, a mixer, and a loop player, as well as a variety of audio effects. The program also included full-featured MIDI sequencing and editing. However, early versions did not support audio tracks and could only play back recorded audio through the sampler, drum machine, or loop player.

From its first version, Reason introduced a unique interface and workflow that set it apart from other DAWs and that remain in place to this day. Many DAWs show devices and plug-ins as a text list and require separate clicks to open each device in a new window. And they typically offer only a limited amount of screen space to represent audio devices. In contrast, Reason presents a virtual studio rack view that shows all of the devices for a project in a single window, with all of the controls for each device immediately available and accessible. (Only VST plug-ins appear in separate windows.)

The studio rack can be "turned around" with a single keypress, revealing virtual jacks on the back of each device. These jacks allow audio and control voltage (CV) signals to be routed to and from the devices in a variety of ways, creating powerful sound design and processing possibilities that are difficult to achieve in other DAWs.

Since the first version, Reason has evolved in the depths of its capabilities and its integrations with other software. Version 6 in 2011 merged the features of another Propellerhead product called Record into Reason. That evolution gave Reason full audio track recording and editing capabilities, as well as a full virtual mixing console. Version 6.5 in 2012 added support for a plug-in format developed by Propellerhead (now Reason Studios) called Rack Extensions, which allows third-party developers to create new devices for the Reason rack that live alongside the built-in devices. In 2017, version 9.5 added support for VST plug-ins, opening up Reason projects to another enormous world of third-party tools. In 2019, version 11 turned the Reason rack itself into a plug-in, allowing Reason's devices to be easily used in other DAWs.

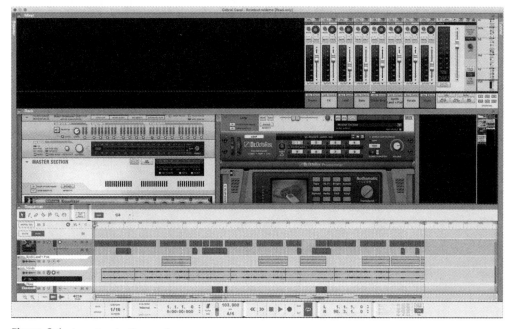

Figure 2.1 A project in Reason Intro

 The website https://www.reasonistas.com/i-use-reason contains an excellent collection of release notes for each version of Reason released over the years, showing the history of its development.

Avid Pro Tools

Pro Tools became the first commercially successful DAW in the audio world when it was introduced back in 1991. The original version of Pro Tools could play only four tracks of audio and cost around $6,000 for the hardware and software.

Since then, Pro Tools has evolved from a tool built exclusively for audio professionals to the popular entry-level DAW that it is today. The Pro Tools audio editing feature set is highly regarded for its power and simplicity, making it commonplace in professional music production and audio post-production for film and TV.

In the 2000s, Pro Tools made great strides as a MIDI production tool, adding a number of advanced MIDI features as well as the ability to edit using standard music notation. Cross-integration with the Sibelius notation platform has made it easy to exchange MIDI information between Pro Tools and Sibelius for sophisticated, professional-grade scoring.

Figure 2.2 A project in Pro Tools | First, the free entry-level version of Pro Tools

Apple GarageBand

GarageBand is Apple's entry-level DAW. As an Apple product, it is available only for Mac OS computers and iOS devices, and it won't run on Windows computers or Android devices.

GarageBand makes it easy to get up and running to start making music right away. This DAW includes a nice complement of effects plug-ins, including EQ, compression, reverb, and delay, as well as excellent guitar amp and pedal emulations.

The included virtual instrument plug-ins span the gamut from synthesizers to acoustic instruments to an automatic drum pattern generator. GarageBand also features a large selection of Apple Loops, which are professionally recorded sound files that can be used to build a song very quickly.

The mobile version of GarageBand can run on both iPads and iPhones, making it easy for users to sketch out a song on the go and then transfer the files to a Mac OS computer for further polishing. You can also import GarageBand songs directly into Apple's professional DAW, Logic Pro X.

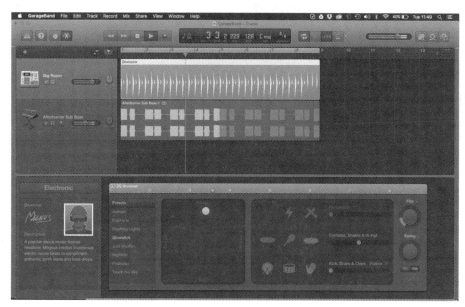

Figure 2.3 The Drummer instrument in Apple GarageBand

Apple Logic Pro X

Logic is GarageBand's big brother. Like GarageBand, Logic runs only on Mac OS and iOS platforms. The product was originally developed by a company called Emagic, which got its start creating MIDI sequencing software for Atari computers.

Logic includes the same core feature set as GarageBand, but it expands upon that to offer a truly professional audio production tool. Logic features a huge number of plug-in effects, including pro-level EQ, dynamics, and modulation effects. It also features a convolution reverb effect that can be used to create ultrarealistic models of real acoustic spaces.

The included virtual instruments in Logic are equally impressive. Logic includes a number of synthesizers, a physical modeling synth, a sampler, several drum instruments, as well as an organ and a clavinet. Many MIDI composers and producers select Logic as their go-to DAW due to its robust support for virtual instruments.

Additional power-user features in Logic include Flex Time and Flex Pitch. These features allow users to freely adjust the timing and pitch of recorded audio.

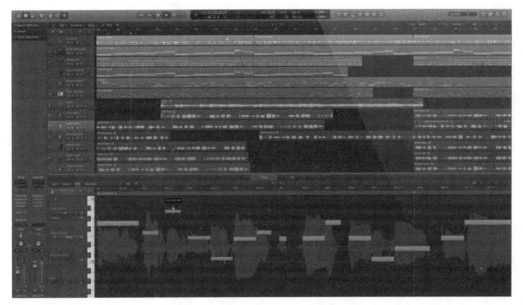

Figure 2.4 Flex Pitch editing in Apple Logic Pro X

 The time stretching feature in Reason provides functionality similar to Logic's Flex Time.

Ableton Live

Ableton's Live is a unique DAW that was created specifically as an electronic music performance tool. It offers a familiar track layout and audio and MIDI editing toolset in its Arrangement view. But where Live really stands out is the revolutionary Session view that took the DAW world by storm back in 2001.

In Session view, audio and MIDI clips are loaded into slots that can be triggered, stopped, and looped independently from one another. As a result, complex song arrangements can be performed on the fly in a way that isn't possible with many other DAWs. This has made Live a favorite among live performers, DJs, and electronic musicians.

The inclusion of the Cycling '74 Max software with Ableton Suite (known as Max for Live) offers another unique feature—the ability to create custom effects and virtual instrument plug-ins without writing any code.

In recent years, Live has probably seen more development of dedicated control surfaces than any other DAW, with Akai's APC series, Novation's Launchpad series, and Ableton's Push devices leading the way.

Figure 2.5 The Session view in Ableton Live

Steinberg Cubase/Nuendo

Steinberg's Cubase is one of the oldest and most respected DAWs on the market. The product began its development in the 1980s as a MIDI sequencer for the Atari ST (like Logic). Cubase has since evolved into a world-class DAW with outstanding MIDI capabilities and solid audio editing.

In particular, Cubase is very popular with hardcore MIDI users (such as film composers). This is due to several innovations that Steinberg introduced, including an ingenious chord track and elegant sampler articulation management. Additionally, the included VST plug-ins are top-notch.

 For information on VST and other plug-in formats, see "Plug-In Formats" later in this chapter.

Steinberg also manufactures another well-known DAW called Nuendo.

Nuendo is essentially an enhanced version of Cubase with more advanced audio features that target the audio post-production industry. Cubase and Nuendo both run on Mac OS and Windows computers, making either option an excellent choice for users who need to work on Windows or move between the two platforms.

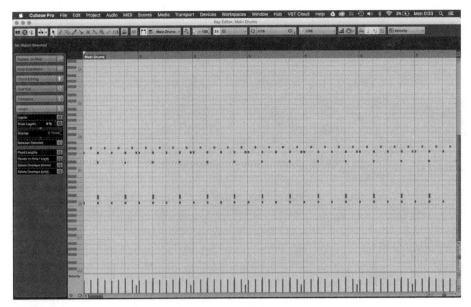

Figure 2.6 The powerful key editor in Steinberg Cubase

And the Rest...

Numerous other DAW options are available, with more coming to market every day. Some examples of other popular DAWs include the following:

- **Image-Line FL Studio**—FL Studio evolved from the Windows-only FruityLoops application that began as a drum machine and sequencer. It now offers powerful audio functionality and has recently crossed over to Mac OS.

- **Cockos REAPER**—REAPER is an inexpensive and flexible DAW. Many users are drawn to REAPER for its incredible customization options: REAPER allows users to create complex actions to accomplish almost any task. In addition, REAPER can be configured so that its editing features mimic the functionality of another DAW, making it easy to use for "switchers."

- **MOTU Digital Performer**—Digital Performer has a loyal following in the film-scoring world. Its credentials include well-known users such as Danny Elfman. Digital Performer offers a number of editing tools that hold particular appeal for classically trained composers and orchestrators, as well as a deep general toolset for MIDI and audio work.

- **PreSonus Studio One**—Studio One is a versatile DAW that has garnered a lot of attention in recent years. It offers a nice balance of MIDI and audio tools, and PreSonus is constantly adding innovative new features such as harmonic editing and components focused on the mastering process.

Plug-In Formats

Most DAWs support plug-ins for both audio processing and MIDI virtual instruments. But not all DAWs support plug-ins in the same format. As a result, a fair amount of confusion surrounds the issue of plug-ins and plug-in formats.

What Is a Plug-In?

A plug-in is a small software program that runs inside of your DAW to extend its functionality. Plug-ins are typically inserted onto a track in the DAW to run in real time. For Reason, devices and plug-ins are added to the rack, and audio is routed via virtual cabling on the back of the devices.

Plug-ins fall into two general categories:

- **Effects Plug-Ins**—Effects plug-ins offer a range of signal processing effects, including equalization (EQ), dynamics processing (such as compressors and limiters), modulation (such as chorus and flange effects), harmonic processing (such as distortion), and time-based effects (such as reverb and delay).

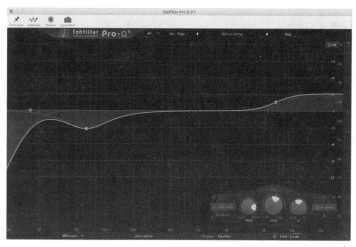

Figure 2.7 FabFilter's Pro-Q 3 is an example of an effects plug-in.

- **Virtual Instrument Plug-Ins**—Virtual instrument devices and plug-ins are software emulations of hardware musical instruments. These instruments include synthesizers (both analog emulations and modern digital algorithms), samplers, and drum machines.

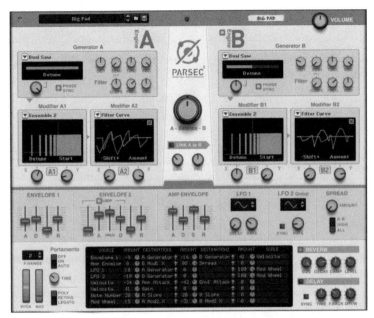

Figure 2.8 Parsec 2 by Reason Studios is an example of a virtual instrument plug-in.

Common Plug-In Formats

Some common plug-in formats include the VST and AU formats developed by Steinberg and Apple, respectively. Reason Studios created the Rack Extension plug-in format specifically for Reason. Avid developed the AAX format for modern Pro Tools systems.

The following breakdown provides details on each of these formats:

- **VST**—Virtual Studio Technology (VST) is a plug-in format created by Steinberg. It was originally developed for Cubase in the 1990s, but the format was rapidly adopted in most of the other DAWs of that era. VST is easily the most popular format for third-party plug-ins and is supported in every major DAW (including Reason) except Logic, GarageBand, and Pro Tools.

- **AU**—The Audio Units (AU) format is Apple's audio plug-in format. AU plug-ins are used primarily in Logic and GarageBand, but they are also supported in several other DAWs (including Ableton Live) as well as a number of simpler audio applications on Mac OS.

- **Rack Extension**—The Rack Extension format is Reason Studios' plug-in format that allows new devices to be added to the Reason rack alongside its built-in devices. Rack Extension devices are fully integrated into the Reason workflow. They support undo/redo, custom virtual cabling jacks, and easy access to automation features. Rack Extensions are distributed only through the Reason Studios online shop, which provides a single centralized location to easily acquire, manage, and install the devices.

- **AAX Native and AAX DSP**—The Avid Audio eXtensions (AAX) specification is the current plug-in format supported in Pro Tools. This specification supports both AAX Native and AAX DSP formats. AAX is the only plug-in format supported in Pro Tools 11 and later. AAX plug-ins are not currently supported in any other (non-Avid) DAW.

Reason Systems

Reason systems typically include a combination of both software and hardware components. With the multitude of options available, it can be difficult to know where to start when building a Reason-based system. The following sections break down some of the choices for you.

Reason Software Options

The Reason software product line for desktop computers includes three options:

- Reason Intro

- Reason

- Reason Suite

> (i) Reason Studios also offers a line of mobile applications for phones and tablets, such as Reason Compact.

Reason Intro is the reduced-cost, feature-limited edition of Reason software. Version 11 of Reason Intro provides the following capabilities:

- Numerous built-in instruments and effects

- Up to a total of 16 tracks (audio or instrument/MIDI)

- Time stretching and pitch editing

- A mixer interface with built-in EQ and dynamics sections on each channel

- Rack Extension and VST plug-in support

- The Blocks song arranging tool

Full Reason software provides unlimited track counts, a complete set of built-in devices, additional sound library content, and support for the ReWire protocol. Reason Suite includes all of the content from the full version of Reason and also bundles in an additional 16 Rack Extension plug-ins for more instrument and effects options.

Table 2.1 provides a feature comparison of basic functionality for the three Reason software options.

Table 2.1 Comparison of Reason software features

FEATURES	Reason Intro	Reason	Reason Suite
Instrument Devices	11	17	28
Effects Devices	11	29	31
Player Devices	3	3	6
Tracks	16	Unlimited	Unlimited
Complete Sound Library	No	Yes	Yes
ReWire Support	No	Yes	Yes
ReFill Sound Library Support	Limited	Full	Full
Mix Channel Send Effects	8	8	8
ASIO, Core Audio Support	Yes	Yes	Yes
Supported Plug-In Formats	Rack Extension, VST	Rack Extension, VST	Rack Extension, VST

Reason Hardware Options

All versions of Reason software support any ASIO-compliant (Windows) or Core Audio–compliant (Mac) audio interface. You have the option to use either your computer's built-in audio hardware or an external interface that connects via USB, FireWire, or Thunderbolt.

Figure 2.9 Focusrite's Scarlett 2i2 is an excellent entry-level interface.

Figure 2.10 Mackie's Onyx Producer 2-2 also works well with Reason.

Figure 2.11 The PreSonus AudioBox USB 96 is another affordable interface option.

Downloading and Installing Reason Intro

If you are ready to proceed with a version of Reason as your DAW of choice, you will need to start by installing the software. Several steps are required for the download and installation process. The steps are illustrated below for Reason Intro. Other editions of Reason software use a similar approach.

Accessing the Installer

To begin the process of downloading Reason Intro, you'll first need to decide how to acquire a copy. Reason software (Intro, standard, and Reason Suite) can be purchased from a number of retailers, as well as directly through the Reason Studios online shop. Prior to purchasing, you can also obtain a free trial version of the standard edition from the Reason Studios website at https://www.reasonstudios.com/en/reason/tryreason.

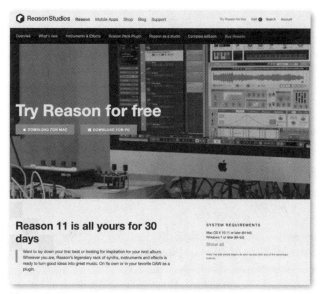

Figure 2.12 The Reason trial download page

Installation Steps

To install Reason Intro, follow these general instructions:

1. **Create or log in to your Reason Studios account**—A Reason Studios account is required to license and activate your Reason products. If you do not already have an account, create one for use with Reason Intro using the form on the Reason Studios website. (See Figure 2.13.)

Figure 2.13 Create a Reason Studios account or log in to your existing account.

2. **Register Reason Intro to your account**—Depending on your purchase method, you may need to register your Reason software to your account. Follow the printed or online instructions that accompanied your purchase, if applicable.

3. **Access the Reason Intro installer**—Before installation, you will need to download the Reason Intro software installer. If you do not complete the download at the time of purchase, you can access the files later by going to the My Products page for your account. (See Figure 2.14.)

My Products

Applications   Rack Extensions   VST Plugins   ReFills   Rent-to-Own

AUTHORIZE   REGISTER NEW PRODUCT

Reason Intro
Version 11

Show details...

DOWNLOAD

Figure 2.14 The My Products page within a user account

From here, click on the **DOWNLOAD** link to access the Reason download page. Then click on the link for the appropriate installer for your system (Mac or Windows).

Figure 2.15 Reason download page

This download may take quite a long time, depending on your Internet connection speed. The installer will be downloaded, packaged inside of a disk image file.

4. **Install Reason Intro**—Finally! In this last step, you will run the installer (from your downloaded disk image file). Double-click on the Reason disk image file to launch it, as needed. Then double-click on the Reason installer file to run the installer. Follow the onscreen prompts to complete the process.

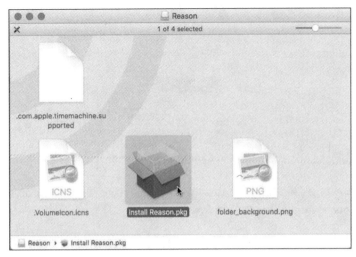

Figure 2.16 The installer file for Reason software (Mac OS shown)

Launching Reason Intro for the First Time

The first time you launch Reason Intro, you will need to specify a few settings. When Reason Intro starts up, it will present the login screen, which allows you to access your copy of the software.

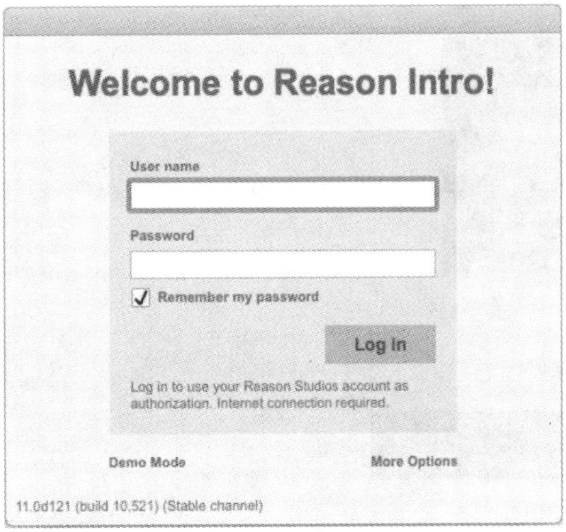

Figure 2.17 Login screen

The login screen presents a few different options to access Reason Intro. The easiest method, and the one you should use initially, is to simply enter the username and password associated with your Reason Studios account and then click the **LOG IN** button.

Accessing Reason Intro by supplying your credentials in this way requires an active Internet connection—not only during the initial login, but continually while working in the software. If you lose your Internet connection while working, you will be able to continue and save your progress, but the software will enter Demo Mode, restricting the functionality of the application. When you reconnect to the Internet, Reason Intro will exit Demo Mode and resume running normally.

 When Reason Intro is running in Demo Mode, an indicator will appear in the transport area. (See Figure 2.18.)

Figure 2.18 Demo Mode indicator

About Demo Mode

Reason Intro may run in Demo Mode under a few different scenarios. This commonly happens when you lose your Internet connection while running a licensed version. However, if you don't own a Reason Intro license, you can also launch into Demo Mode from the login screen to try out the software. You may also encounter Demo Mode if you have a trial license of Reason that has expired.

Demo Mode adds a number of restrictions to using the software. Most notably, you cannot open project files other than demo songs in this mode. If you already had a song open, you can continue working and saving that song and you can start new, empty songs to try out ideas, but you cannot open any other projects you have previously saved. You also cannot bounce or export audio from Reason Intro in Demo Mode, so you can't take a song out of Reason Intro to work on it elsewhere. Finally, most Rack Extension devices will not be available in Demo Mode.

To avoid having to enter your username and password upon launch, and to be able work offline without requiring an Internet connection, you can install your Reason Intro license to your computer (or to a USB device called an Ignition Key). This lets you authorize Reason Intro with the installed license so that the login process is not required.

 To learn more about authorizing to a computer or Ignition Key, click the More Options link on the login screen and follow the onscreen instructions.

Once you have logged into Reason Intro for the first time, the application will take a moment to scan for VST plug-ins installed on your system. If you don't have any VST plug-ins, you likely won't notice this process at all. If you have many plug-ins on your computer, the initial scan may take some time, and you will see a progress window while Reason Intro is working.

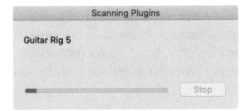

Figure 2.19 Scanning Plugins progress window

Afterwards, the Reason Intro Setup Wizard will appear and walk you through a few configuration steps to get fully up and running. First, on Windows computers, Reason Intro will present an option to choose the language used in the software. Next and most importantly, Reason Intro will ask you to configure your audio settings.

Select your audio device or driver from the list provided in the window. You also have the option to choose a sample rate; you can leave this value at the default of 44,100 (44.1 kHz) to start out. Then click the **NEXT** button at the bottom of the dialog box.

 You can change both your audio device and your sample rate at any time in Reason Intro's preferences.

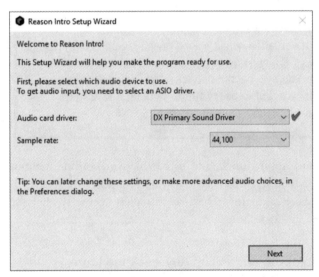

Figure 2.20 Selecting an audio device and sample rate in the Reason Intro Setup Wizard (Windows)

After you've configured the audio settings, Reason Intro will scan your system for musical keyboards and other controllers that are attached to your computer. Reason Intro provides built-in support for a large number of different music control hardware devices. If you have supported hardware attached, the Reason Remote system will automatically map the buttons, knobs, faders, and other controls on your hardware to the virtual controls of Reason Intro devices. You can also use controllers and MIDI keyboards with Reason Intro that don't have built-in Remote support for mapping hardware controls.

Click the **NEXT** button when the scanning process completes to continue the setup process.

 You can add MIDI devices at any time in Reason Intro's preferences.

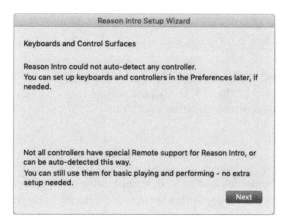

Figure 2.21 Reason Setup Wizard keyboard and control surface detection

The last step in the Setup Wizard lets you opt in to send usage statistics and error reports to Reason Studios. You can set this option to your preference and then click **FINISH** to proceed.

With the Setup Wizard complete, the main Reason Intro interface will appear.

You may be prompted by the Manage Content dialog box at this point, which allows you to install optional content that comes with your software. Select the content to install and then click the **INSTALL** button; to install all content, simply click the **INSTALL ALL** button.

Important Concepts in Reason

There are a few key DAW concepts that you should master before getting started with audio software such as Reason. For starters, you should understand how the software handles audio data for your projects. Additionally, you should recognize the fundamental differences between audio and MIDI data.

Song Files and Audio Data

Reason manages one main file type, known as a song file. When you first start working on a new project in Reason, the project will not yet be stored on your hard drive. Once you use the Save command for a new project, Reason will save a song file with a name and location that you specify. A song file contains all of the information about the current audio production. This information includes the number and type of tracks in the project, the position of audio and MIDI data on each track, the devices used in the rack, the mixer level settings, the project settings, and so on. You will typically create a unique song file for each song or production you work on.

Unlike some DAWs, Reason does not store separate audio files representing audio recordings that you've created in your project or imported into the project from a disk location. Instead, Reason stores recorded or imported audio data within the song file you are working on. This way, you never have to worry about your audio files becoming separated from your song files, and you only need to keep track of one file for each project. However, because audio is embedded in the song files, projects with many recorded tracks or a large amount of imported audio can occupy a large amount of disk space.

 Because song files contain audio data, you may want to be careful when using the Save As command to create a new copy of a song file. Doing so will effectively copy and duplicate all of the included audio data as well.

Audio Versus MIDI

The differences between audio and MIDI data are very important to understand when you're just starting out in the world of audio production. Audio files represent the audio signal that is recorded when the sound (from a voice, instrument, or other source) is captured using an audio interface. Therefore, the audio file itself actually *contains* the sound information. Audio files can be played back through the audio interface, allowing the sound to be heard using monitor speakers or headphones.

On the other hand, MIDI data is a series of note and control messages, typically recorded from a MIDI controller (like a keyboard or drum pads). MIDI data can also be manually entered in Reason.

The MIDI files you create do *not* actually contain any sound information. A MIDI file's note and control messages must be sent to a virtual instrument or sound module, which creates the associated audio signal.

 Reason uses the terms *note events* and *note clips* to refer to the groups of note and control messages used in songs. However, this data is functionally the same as MIDI data, as described above, and represents a series of note and control messages that a virtual instrument can turn into an audio signal.

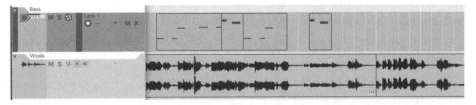

Figure 2.22 A Reason instrument track with note clips (top) and an audio track with audio clips (bottom)

Review/Discussion Questions

1. What are some tasks that can be completed using a digital audio workstation (DAW)? (See "Functions of a DAW" beginning on page 32.)

2. What are some of the differences between the common DAWs? How are they similar? (See "Common DAWs" beginning on page 32.)

3. What is an audio plug-in? How are plug-ins enabled in a DAW? (See "What Is a Plug-In" beginning on page 39.)

4. What are the two general categories of plug-ins? Give some examples of plug-ins in each category. (See "What Is a Plug-In" beginning on page 39.)

5. Which of the common plug-in formats is/are supported in Reason Intro? (See "What Is a Plug-In" beginning on page 39.)

6. What are the three versions of Reason software? (See "Reason Systems" beginning on page 41.)

7. What are some of the differences between Reason Intro, standard Reason software, and Reason Suite? (See "Reason Systems" beginning on page 41.)

8. What are some examples of audio interfaces that are supported in Reason Intro? (See "Reason Hardware Options" beginning on page 42.)

9. What are the general steps required to install Reason Intro? (See "Downloading and Installing Reason Intro" beginning on page 43.)

10. What options are available to access Reason Intro from the login screen? (See "Launching Reason Intro for the First Time" beginning on page 46.)

11. What is Demo Mode in Reason Intro? Describe some situations that might cause Reason Intro to run in Demo Mode. (See "Launching Reason Intro for the First Time" beginning on page 46.)

12. What are some of the settings you'll need to specify upon first launching Reason Intro? (See "Launching Reason Intro for the First Time" beginning on page 46.)

13. What is the primary type of file that Reason manages? (See "Important Concepts in Reason" beginning on page 49.)

14. What are some of the differences between audio and MIDI information? (See "Audio Versus MIDI" beginning on page 50.)

 To review additional material from this chapter and prepare for certification, see the Reason Audio Production Basics Study Guide module available through the Elements|ED online learning platform at ElementsED.com.

Setting Up a Multitrack Project

🎧 Activity

In this exercise, you will create a simple project based on a remix of the song "Overboard" by The Pinder Brothers. You will then configure the project settings for playback and play through the project.

🕐 Duration

This exercise should take approximately 10 minutes to complete.

⊕ Goals/Targets

- Launch your Reason software

- Set the correct song tempo

- Import audio to the project

- Set preferences to configure the project for playback

Exercise Media

This exercise uses media files taken from the song, "Overboard," provided courtesy of Sacramento-area band The Pinder Brothers, along with alternate files provided by Eric Kuehnl from a synthesizer-based remix of the song.

Written by: Matt Pinder; Performed by: The Pinder Brothers;
Produced by: Scott Reams and The Pinder Brothers*; Remix by: Eric Kuehnl and Frank D. Cook

The media provided for this course may be used for educational purposes only. No rights are granted to use the media for any other personal, commercial, or non-commercial purposes.

* The mix, processing, and media files have been adapted for use in the exercises contained herein.

Getting Started

To get started, you will launch your Reason software as described in Chapter 1. The exercises in this book can be completed using Reason Intro, the standard full version of Reason, or Reason Suite.

 Before starting this exercise, Reason software must be installed on your system. If necessary, complete the download and installation steps outlined in Chapter 2.

Launch your Reason software:

1. Start the application by doing one of the following (or using another method of your choice):

    • On a Mac-based system, click on the **REASON** icon in the Dock. Or, click the **LAUNCHPAD** icon, type "Reason," and click on the application icon.

    • On a Windows-based system, double-click on the **REASON** shortcut on the desktop. Or, press the **START** key, type "Reason," and click on the application icon.

2. Log in to your Reason Studios account as needed.

A new song will be created and will display on screen.

Before continuing on to import the tracks for the song, you will need to set the correct song tempo. We will be using a tempo of 116 beats per minute (BPM).

Set the song tempo:

1. Locate the tempo value in the Transport Panel at the bottom of the screen (below the main window):

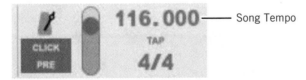

Figure 2.23 The song tempo in the Transport Panel

2. Do one of the following:

    • Click on the number and drag it up or down.

    • Click on the number and type **116** in the highlighted field; then press **ENTER** or **RETURN** to set the tempo.

Setting the tempo before importing the audio is important to ensure that the beats of the song line up with the musical grid in Reason. Although you can change the tempo after audio has been imported, Reason will try to time-stretch the audio to follow the tempo change, which is not intended for this exercise.

 In many real-world situations, you may receive audio with an unknown tempo, an inconsistent tempo, or tempo changes. In these cases, more complex techniques are needed to determine the tempo or to map the song's tempo to the audio.

Importing Audio for the Project

In this section of the exercise, you will display the Sequencer view and import audio to the project. The process you'll use will automatically create tracks for the imported audio files.

Show the Sequencer view and open the Browser:

1. Choose **WINDOW > VIEW SEQUENCER** or press **F7**, if needed, to show the Sequencer view.

2. Press **SHIFT+RETURN** (Mac) or **SHIFT+ENTER** (Windows) two times to ensure that the Song Position Pointer is at the beginning of the song.

    This is the equivalent of pressing the Stop button in the Transport Panel. The first Stop command while the transport is stationary returns the Song Position Pointer to the previous playback point; the second returns to the song start.

3. Choose **FILE > IMPORT AUDIO FILE**. The Browser will display (if not already visible), set to the Import Audio File context and pointed to your Documents folder by default or the last folder you visited.

    If you saved the **Reason APB Media Files** folder to a location other than your Documents folder, or if your Browser is not pointed to the Documents folder, you will need to navigate within the Browser to the proper location. (See Figure 2.24.)

    You can quickly navigate to common locations such as the desktop using the **LOCATIONS AND FAVORITES LIST** on the left side of the Browser. You can also use the **ROOT FOLDER** drop-down list to move upwards in the file structure on your computer. Similarly, you can click the **UP ONE LEVEL** arrow button to navigate to the parent folder of the current location.

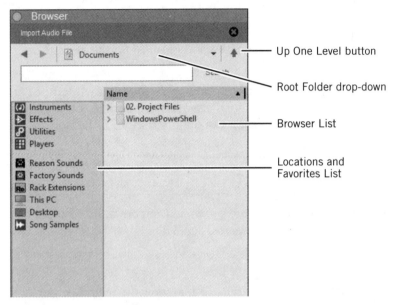

Figure 2.24 The Browser in the Import Audio File context (Windows system pictured)

The Browser List displays the files and folders within the currently selected folder. You can open a folder from the list by double-clicking on it.

4.   Use the browser tools to navigate to the **REASON APB MEDIA FILES** folder; then double-click to open the folder in the Browser List.

Import the project audio:

1.   Open the 02. Project Files folder within the Reason APB Media Files folder by double-clicking it.

2.   Select the twelve included .wav audio files. Reason will start auto-auditioning the files in the Browser.

3.   Click the Stop button (square icon) or the Auto button at the bottom of the Browser to cancel the audition.

4.   Click the Sort Arrow at the top of the file list to display the files in descending numerical order. (See Figure 2.25.)

Figure 2.25 Files selected in the Browser

5. Click **IMPORT** at the bottom of the Browser to complete the process. The imported audio will be placed on twelve new tracks, starting from the bottom of the list.

Playing the Project

Set up the project for playback:

1. Choose **REASON > PREFERENCES** (Mac) or **EDIT > PREFERENCES** (Windows) to display the Preferences window.

2. Click on the **AUDIO** tab at the top of the Preferences window.

3. Verify that your connected audio interface is displayed in the Audio Device menu (Mac) or Audio Card Driver menu (Windows). If needed, click on the menu selector to choose the desired interface or driver.

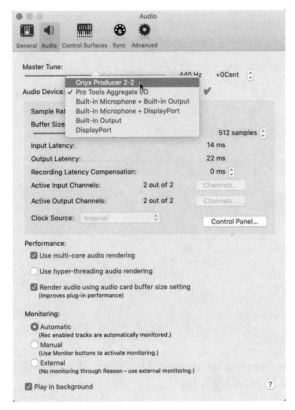

Figure 2.26 Selecting the Onyx Producer 2-2 audio interface as the audio device (Mac system pictured)

4. Close the Preferences window when finished.

5. If you have speakers or headphones connected to your system, make sure the volume control is set to a safe, low level initially.

6. Turn off the Loop function by clicking the **LOOP ON/OFF** button in the Transport Panel at the bottom of the screen so that it becomes disabled (unshaded).

Figure 2.27 Disabling the Loop On/Off button in the Transport Panel

 You can also toggle the Loop function on/off by pressing the L key on your computer keyboard.

Play back the project:

1. Press the **SPACEBAR** to begin playback. The playback cursor will scroll across the screen and you will hear playback through your connected speakers or headphones.

2. While playing back the project, scroll the Sequencer view vertically as needed to view the audio waveforms on each track. You can scroll using the scroll bar on the right side of the window, a scroll wheel on your mouse, a scroll gesture on a trackpad, or similar method.

3. Press the **SPACEBAR** some time after bar 10 or so to stop playback.

4. Press the **SPACEBAR** again to restart playback. Notice that playback picks up from where you had stopped.

5. Stop playback again after a few bars.

6. With playback stopped, press **SHIFT+RETURN** (Mac) or **SHIFT+ENTER** (Windows) or click the **STOP** button in the Transport Panel. The Song Position Pointer will return to the previous playback point.

 Reason includes a preference setting that allows playback to start over from the same position each time. To enable this behavior, select Return to Last Start Position on Stop under Preferences > General.

7. Begin playback again to hear the result.

8. Allow playback to continue through the end of the song snippet; notice that the page scrolls each time the Song Position Pointer reaches the end of the screen.

   This step assumes the Follow Song option is enabled (**OPTIONS > FOLLOW SONG**).

 The Follow Song option and other scrolling behaviors are covered in Chapter 5 of this book.

9. Press the **SPACEBAR** when finished.

Finishing Up

To complete this exercise, you will need to save your work and close the song. You will be reusing this song in Exercise 5, so it's important to save the work you've done.

Finish your work:

1. Press **SHIFT+RETURN** (Mac) or **SHIFT+ENTER** (Windows) two times or click the **STOP** button twice in the Transport Panel to return the Song Position Pointer to the song start. The song will scroll with each action to show the new Song Position Pointer location.

2. Choose **FILE > SAVE** to save the song. A Save dialog box will display, allowing you to specify the song name and location.

3. Name the file Overboard01-*xxx*, where *xxx* is your initials, and choose an appropriate save location.

Figure 2.28 Specifying a name and save location in the dialog box (Mac system shown)

(i) On the Mac, you may need to click the down arrow next to the Save As field to expand the dialog box to select a folder within a selected drive location.

4. Click the **SAVE** button to store the file at the chosen location. A progress bar will appear while the audio files are copied into the song file.

5. When the save is complete, choose **FILE > CLOSE** to close the song.

(i) You can also close a project by closing the main Song document window that contains the Sequencer view.

6. If desired, you can exit Reason by pressing **COMMAND+Q** (Mac) or **ALT+F4** (Windows).

(i) When you close the last open Song document on Windows, the Reason application will automatically exit.

That completes this exercise.

Chapter 3

Audio Recording Concepts

...*What You Need to Record Audio*...

This chapter introduces you to the fundamentals of audio recording. We begin with a discussion on the basics of sound, including the principles of frequency and amplitude. We then discuss various types of microphones and their uses, before diving into a discussion on multitrack recording. We follow this by examining the process of converting audio signals between the analog and digital domains and exploring the role of the audio interface in this exchange. The characteristics of analog and digital audio discussed in this chapter are important considerations when it comes to optimizing your results with any DAW.

⊕ Learning Targets for This Chapter

- Understand the audio characteristics of frequency and amplitude

- Understand characteristics of traditional microphones and USB microphones

- Understand the purpose of multitrack recording

- Understand the basic principles of analog-to-digital conversion

- Recognize the role of the audio interface in modern digital recording

 Key topics from this chapter are illustrated in the Reason Audio Production Basics Study Guide module available through the Elements|ED online learning platform. Sign up at ElementsED.com.

The process of recording audio has been commonplace for more than 60 years, ever since magnetic tape–based recording processes began replacing earlier processes of inscribing a signal to a physical surface. In just the past 20 years or so, tape-based recording has itself largely given way to digital recording technology.

Regardless of the technology used to capture an audio recording, certain fundamental characteristics remain unchanged, including how an audio event moves through an acoustic environment and how that event is translated into an electrical signal that can be recorded. Recording in digital simply adds a dimension of collecting discrete measurements of the signal and storing those measurements as binary information.

The Basics of Audio

To understand the process of recording audio, it helps to first understand the basics of audio and sound waves. Sound waves are created when a physical object vibrates, causing a variation in the surrounding air pressure. By way of example, consider what happens when a guitar string is plucked.

The vibration of the guitar string, as it moves back and forth in a repeating cycle, causes a displacement of air particles. This results in cyclical variations in air pressure, known as compression and rarefaction. The air pressure increases (compression) and decreases (rarefaction) in a way that corresponds directly to the pattern and frequency of the string's vibration. This cycling pattern of air pressure is referred to as a sound wave.

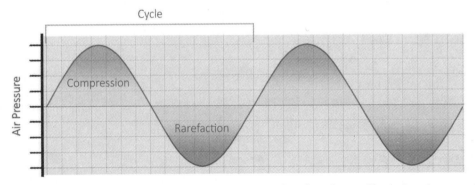

Figure 3.1 Cycles of compression (increasing air pressure) and rarefaction (decreasing air pressure) in a sound wave

Frequency

The frequency of the cycle of sound pressure variations determines our perception of the pitch of the sound. If the guitar string vibrates quickly, the air pressure variations will likewise cycle quickly between compression and rarefaction, creating a high-pitched sound. If the guitar string vibrates slowly, the air pressure will cycle slowly, creating a low-pitched sound.

We measure the frequency of these changes in cycles per second (CPS), also commonly denoted as hertz (Hz). These two terms are synonymous—15,000 CPS is the same as 15,000 Hz. Multiples of 1,000 Hz are often denoted as kilohertz (kHz). Therefore, 15,000 Hz is also written as 15 kHz.

The open A string on a guitar vibrates at a frequency of 110 Hz in standard tuning. Playing the A note on the 12th fret produces vibrations at 220 Hz (one octave higher).

The range of human hearing is generally accepted to be between 20 and 20,000 cycles per second. This is also commonly denoted as 20 Hz to 20 kHz. Therefore, to capture a full-spectrum recording, we must be able to represent all frequencies within this range.

Amplitude

The intensity of a sound, or the amount of change in air pressure it produces, creates our perception of the loudness of the sound. A sound's intensity is represented by the amplitude (height) of the sound wave. We measure amplitude in decibels (dB).

The decibel scale is defined by the dynamic range of human hearing, from 0 dB at the threshold of hearing to approximately 120 dB at the threshold of pain. The decibel is a logarithmic unit that is used to describe a ratio of sound pressure. As such, the decibel does not have a linear relation to our perception of loudness. An increase of approximately 10 dB is required to produce a perceived doubling of loudness.

By way of example, the amplitude of ordinary conversation is around 60 dB. Increasing the amplitude to around 70 dB would essentially double the loudness, similar to what you might face trying to hold a conversation in a room with the TV or radio on. Increasing the amplitude to 80 dB would double the loudness again, such as you might experience trying to hold a conversation in a crowded room.

Microphones

The most common technique used to capture an acoustic audio event for recording is to place a microphone somewhere near the sound source. Microphones come in many shapes, sizes, and prices, so it can help to know a little bit about how they work before selecting the microphone or microphones you plan to use.

Traditional Microphones

All microphones are transducers: they change energy from one form to another. A traditional analog microphone simply converts the variations in air pressure into variations in an electrical current.

Three basic types of microphones are commonly available to make this conversion: dynamic mikes, ribbon mikes, and condenser mikes.

Dynamic Microphones

A dynamic microphone consists of a diaphragm attached to a coil of wire encircling a magnet. When sound waves hit the diaphragm, it vibrates back and forth, thereby moving the coil of wire back and forth over the magnet. This movement induces a current in the coil, with compression creating a positive voltage and rarefaction creating a negative voltage.

The advantages of dynamic microphones are that they are rugged and durable. They can also handle high input signals without distortion. Dynamic microphones are commonly used in both studio and live performance environments.

Ribbon Microphones

A ribbon microphone consists of a thin metal foil or other conductive ribbon material suspended between the positive and negative poles of a magnet. As sound waves hit the ribbon, they cause it to vibrate; the ribbon's movement within the magnetic field induces a current within the ribbon itself.

The advantages of ribbon microphones include their warm, smooth tone and their ability to capture high-frequency detail. Some ribbon microphones can be delicate and expensive. Ribbon mikes tend to be better suited for use in a studio environment rather than live performance.

Condenser Microphones

Condenser microphones consist of a thin conductive diaphragm (front plate) affixed a small distance in front of a metal back plate. The two plates are typically energized with a fixed charge. When sound waves hit the front plate, the diaphragm vibrates, varying the distance between the two plates. This varies the capacitance of the circuit and creates an opposite change in electrical voltage.

 The capacitance is the ratio of the electric charge on the plates to the voltage difference between them. The capacitance increases as the plates move closer together and decreases as the plates move farther apart.

Condenser microphones require power to operate, either from a battery or from a phantom power source. Phantom power is typically 48 volts DC applied through a microphone cable to the condenser mike.

 Mixers, microphone preamps, and audio interfaces commonly provide the option to enable phantom power, often denoted as +48V on the device.

The advantages of condenser microphones include a wide, smooth frequency response and sharp, detailed transient response. Condenser mikes are particularly well suited for acoustic instruments and cymbals. Although condenser mikes are very popular for in-studio recording, they are also commonly used in live sound production, particularly as drum overhead mikes.

USB Microphones

Another option to consider for home or studio use with your DAW is the USB microphone. USB mikes function the same as their traditional microphone counterparts, in that they convert acoustical energy into electrical energy. However, a USB mike goes a step further using two additional circuits: a built-in preamp and an analog-to-digital converter. This additional circuitry allows the microphone to be plugged directly into your computer for use with your DAW without requiring a separate audio interface.

For information on the analog-to-digital conversion process, see "Moving Audio from Analog to Digital" later in this chapter.

Most USB microphones are condenser mikes, although a few dynamic USB mikes are also available.

Popular choices among USB microphones include:

- **Blue Microphones**

  - **Snowball USB Microphone**—Designed for podcasting and other desktop recording applications ($$)

  - **Yeti USB Microphone**—Designed for studio vocals, musical instruments, voiceovers, field recordings, podcasting, and desktop recording ($$$)

  - **Spark Digital Lightning Microphone**—Designed as a premium price option for studio and desktop recording for voiceovers, vocals, and instruments ($$$$)

- **CAD U37 USB Studio Condenser Recording Microphone**—Designed as a budget option for desktop recording and voiceovers, vocals, and instrumental recording ($)

- **Audio-Technica ATR2100-USB Cardioid Dynamic USB/XLR Microphone**—Designed as a mid-price mike for stage and studio use ($$)

- **RØDE NT-USB USB Condenser Microphone**—Designed for studio-quality recording of vocals, instruments, and voiceovers with built-in monitoring control ($$$)

- **Apogee Mic 96k Professional Quality Microphone**—Designed as a premium price option for high-resolution studio-quality recording ($$$$)

- **RØDE Podcaster USB Dynamic Microphone**—Designed as a premium price option for studio-quality podcasting and other desktop recording applications ($$$$)

Other Considerations

Aside from the characteristics of different types of microphones described above, a few other considerations should come into play when selecting the right mike for your needs.

Polar Pattern

A microphone's polar pattern describes how the microphone responds to sound coming from different directions. Typical choices include omnidirectional mikes, bidirectional mikes, and unidirectional mikes.

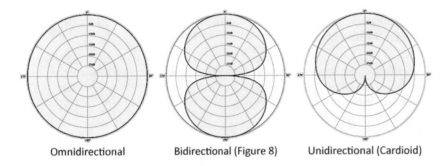

Omnidirectional   Bidirectional (Figure 8)   Unidirectional (Cardioid)

Figure 3.2 Common polar patterns: omnidirectional, bidirectional, and unidirectional

Omnidirectional. An omnidirectional mike will pick up sound equally from all directions. While this may be desirable for a microphone placed in the middle of a large conference table, it is usually not the best choice for studio recording where the goal is to isolate a sound source.

Bidirectional. Bidirectional mikes pick up sound primarily from the front and rear of the microphone, rejecting sound coming from the sides. This kind of mike may be desirable for across-the-desk interviews, two-part vocal recordings, or similar recordings of two opposite-facing sound sources. By rejecting sound from the sides, this type of mike will limit the amount of room acoustics and unwanted ambient noise in the recording.

Unidirectional (cardioid) . Unidirectional mikes are the most popular choice for recording isolated sound sources. These are also commonly referred to as cardioid microphones. Unidirectional, or cardioid, polar patterns are designed to pick up sound from the front of the microphone, while rejecting (or attenuating) sound coming from the sides or rear of the microphone.

Proximity Effect

Most unidirectional and bidirectional microphones exhibit a noticeable bass boost when used within a few inches of a target sound source. This phenomenon is known as the proximity effect and often results in an unwanted boomy sound. You can hear the effect by speaking into a microphone as you move up close to the front of the grille.

If you are aware of a mike's proximity effect, you can adjust the mike placement or roll off the bass frequencies with an equalizer to compensate.

Basic Miking Techniques

Microphone placement and position have a large impact on how well a microphone will perform at capturing the desired signal. The overall goal when setting up a microphone is to target the sounds you want to record while rejecting the sounds you don't. Several accessories are available to help with this process.

Microphone Accessories

The microphone accessories discussed below can assist you with mike placement, noise reduction, and sound isolation.

- **Mike Stand**—USB microphones designed for desktop use often come with a small stand or clip. For miking in a studio or an open environment, however, you will need to invest in a full-sized mike stand. You might also consider a stand with a boom extension or gooseneck add-on for better reach and flexibility. You'll find a variety of stands and extensions readily available from your local music store or online retailer.

- **Windscreens and Pop Filters**—Windscreens and pop filters are designed to reduce unwanted noise resulting from wind or bursts of air hitting the microphone. These problems are especially prevalent for on-location field recordings and close-miked recordings of vocal or dialogue performances.

  > (i) In vocal performances, bursts of air are commonly produced from words that begin with "p" or "b" sounds. These bursts, or plosives, can ruin a recording if they hit a microphone's diaphragm directly.

- **Shock Mounts**—Shock mounts are designed to suspend a microphone, isolating it from any mechanical vibrations that may affect the mike stand. Such vibrations are commonly caused by loud music, noise carried through a floor or stage, or nearby thumps and bumps.

- **Acoustic Shields and Vocal Booths**—To improve sound isolation, especially for vocal and dialogue recordings, you can consider using an acoustic shield. These products are designed to reduce the influence of ambient noise, room acoustics, and reflected sound on a recording. Acoustic shields are most effective at attenuating mid- and high-frequency audio.

  A more extreme approach is to build or install a fully enclosed vocal booth or isolation room. This will be more effective at reducing unwanted noise than a shield, but it can also be expensive and complex to design.

  Note, also, that both shields and booths may colorize the audio, boosting or cutting certain frequencies of the performance. You may be able to correct for the colorization using an equalizer; however, you will need to determine whether the isolation gains are worth the trade-off.

Where to Shop for Audio Gear

While locally owned music stores are becoming scarce, audio gear has never been easier to shop for. In addition to ubiquitous Guitar Center storefronts, numerous online retail outlets are available. Online options include guitarcenter.com, sweetwater.com, musiciansfriend.com, and vintageking.com, among others. Almost any audio equipment you are looking for can be found with a quick Internet search!

Basic Setup and Connections

Setting up your microphone or microphones is a simple matter of finding the right mike position and establishing a connection to your computer.

- **Microphone Position**—The location of your microphone during a recording can have a dramatic influence on the recorded signal. Generally speaking, the closer the mike is to the target sound source, the cleaner the recording will be. That is to say, the recording will include more of the desired sound and less ambient noise.

  However, as noted above, certain mikes exhibit a proximity effect when placed close to their sound source. This can colorize the sound in ways that may be unwanted, depending on the situation.

- **Microphone Connection**—The way that your microphone connects to your computer will vary, depending on the type of mike you are using. Most analog microphones use a cable with XLR connectors (3-pin) to attach to your audio interface. The audio interface will in turn connect to your computer using a USB cable or other common digital computer connection.

Figure 3.3 Microphone XLR connection to an Mbox Pro audio interface

A digital USB microphone will connect directly to your computer, with no audio interface required.

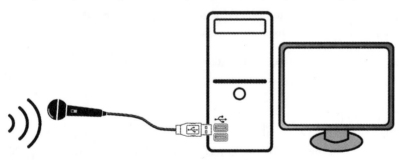

Figure 3.4 USB microphone connection to a desktop computer

Multitracking and Signal Flow

The benefits of recording to a DAW are many. For starters, it allows you to record multiple takes and keep the best parts of each take. Additionally, it gives you the ability to edit and process a recording to improve the final results. But one of the greatest advantages is the freedom to record numerous isolated parts, each on its own track (known as multitrack recording).

What Is Multitrack Recording?

The concept of multitrack recording has been around for many decades and is a fundamental aspect of audio production. Using this production method, each audio source is isolated as an individual recording on its own track. Each of these recorded tracks can then be edited and processed discretely. By playing back all of the individual tracks simultaneously, you can create a mix of the multiple sound sources as a final output.

By way of example, a music recording might include individual tracks for vocals, guitar, bass guitar, keyboards, and drums. By isolating each part on its own track, the volume levels, pan settings, EQ settings, and other processing can be set independently for each component.

Another advantage of multitrack recording is that individual parts can be recorded at different times. This is key if the performers cannot all be present in the same location at the same time. It also allows individual contributors to record multiple parts. For example, a musician can record both guitar and piano parts for a song. Or a voice actor can record dialogue lines for multiple characters in a script for an animated short.

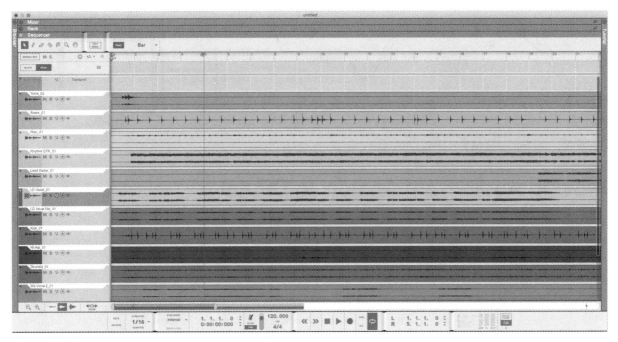

Figure 3.5 Multitrack music production project in Reason Intro

Not all recordings require multiple tracks. In some cases, you may want to record a mix of multiple sources to a single track. Some examples might include recording a live ensemble performance with a pair of microphones routed to a single stereo track or using an external mixer to sum multiple drum microphones to a stereo track. One reason for doing this is to simplify the recording and the number of tracks in your session. You might also use this approach if you do not have enough inputs on your system to record from many microphones at the same time.

Recording Signal Flow

Recording onto a computer involves routing a signal through various processing stages. This routing path is commonly referred to as the signal flow. Here, we will focus on a typical recording signal flow, from the sound source to the storage device (HDD or SSD).

The sound originates at a source, such as a guitar, and travels through an environment by way of variations in air pressure, as discussed above. The air pressure changes are picked up by a microphone, converting the sound into an electrical signal. The electrical signal next travels down the microphone cable to an audio interface.

Within the audio interface, the signal is boosted, using a preamplifier (or *preamp*). The preamp is used to create a healthy signal level for recording. Next, the signal is sampled, using periodic measurements of the electrical voltage. The measured values are represented as strings of binary numbers and are passed along to the computer. At the computer, the binary numbers are stored on a hard disk drive or solid-state drive.

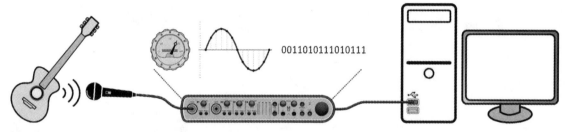

Figure 3.6 Recording signal flow from a guitar to a computer's hard drive

Moving Audio from Analog to Digital

As mentioned above, the practice of recording audio to a computer involves converting an electrical signal into individual, discrete binary measurements. This is the process known as analog-to-digital conversion.

Analog Versus Digital Audio

The variations in electrical voltage produced by a microphone represent a continuously changing analog audio waveform. To represent that information on a computer, the waveform must be converted into binary numbers that the computer can understand.

What Is Binary Data?

Binary data is comprised of binary digits, or bits. Each binary digit can represent only two possible values: zero or one. (At the most basic level, computers store and work with values using switches that are either off or on. A bit value of zero can be stored by toggling the associated switch off, while a bit value of one can be stored by toggling the switch on.)

To store values larger than one, computers use strings of multiple binary digits. For example, using a string of four binary digits, a computer can store values from zero (0000) to fifteen (1111). Using a string of eight binary digits, a computer can store values from zero up to 255. Using a 16-bit string, the computer can store values up to 65,535.

Once an audio signal has been converted to digital, the waveform can be stored, read, and manipulated by the computer. The digital audio file is comprised of individual measured values (samples) stored as strings of binary digits.

The Analog-to-Digital Conversion Process

Converting an analog signal to digital involves two critical parameters: the sample rate and the bit depth.

Sample Rate. The sample rate refers to the frequency at which the incoming electrical signal is measured by the audio interface. The sample rate must be high enough to accurately represent the original analog signal.

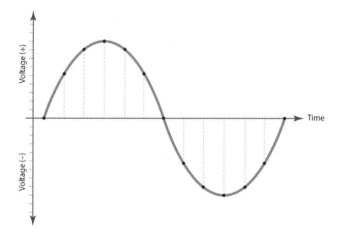

Figure 3.7 Sample intervals used to measure the changing voltage of an incoming audio signal

The sample rate required to accurately record a given signal is driven by a fundamental law of analog-to-digital conversion, known as the *Nyquist Theorem*. The Nyquist Theorem states that, in order to represent a given audio frequency, each cycle of the waveform must be sampled at least two times. Or, stated another way, the sample rate must be at least two times the frequency of the audio you wish to record.

Because we generally want to record full-spectrum audio—from 20 Hz to 20,000 Hz—the Nyquist Theorem tells us we need to use a sample rate of at least 40,000 Hz (or 40 kHz): twice the upper end of this range. Most modern digital audio systems use a minimum sample rate of 44.1 kHz.

Bit Depth. The bit depth (or word length) refers to the number of binary digits included in each measurement string. The more binary digits used, the more accurate the measurements will be. Low bit-depth recordings exhibit a loss of audio detail—the result of a reduced dynamic range.

What Is Dynamic Range?

Dynamic range is the difference between the quietest signal that a system can represent and the loudest signal that the system can represent. Dynamic range is measured in decibels (dB). This is a relative measurement, not an absolute loudness value. The maximum absolute loudness of a system depends on the amount of amplification applied at the output; however, the dynamic range remains constant at all loudness levels, since it measures the relative difference between two levels.

You can estimate the dynamic range of a system by multiplying the bit depth by six. Said another way, each binary digit that is added to the word length provides approximately 6 dB of dynamic range.

The useful range of loudness values for speech and music is generally considered to be from 40 dB on the quiet end to 105 dB on the loud end: a 65-dB dynamic range. To produce that dynamic range, a digital audio system must provide a minimum 11-bit word length (65 dB divided by 6 dB per bit). Additional dynamic range is required to allow for headroom before clipping and to accommodate any inherent noise floor in the system. Most modern digital audio systems use a minimum bit depth of 16 bits.

The Audio Interface

As mentioned previously, an audio interface is a device used to provide analog-to-digital conversion for the audio coming into a system. It also provides digital-to-analog conversion for the audio playing out of the system. The audio interface is what allows you to get sound into and out of your computer.

Most audio interfaces also provide one or more microphone preamps. Preamps generally include a gain control that allows you to adjust the incoming signal and optimize record levels.

Audio Interface Considerations

A multitude of audio interfaces are available today, providing an enormous array of options. To narrow down your choices, it helps to decide on the characteristics that are important for your recording needs.

Some factors to consider include:

- What kind of input and output (I/O) connectivity do you need?
- What level of sound quality are you looking for?
- What sort of budget are you working with?

I/O Connectivity

If you plan to work exclusively in Reason Intro, you could consider an audio interface with multiple inputs. Reason Intro supports hardware with up to 64 inputs; however, the Intro edition is limited to 16 tracks. To record onto all 16 tracks simultaneously, you could theoretically use up to 32 interface inputs (one each for mono tracks or two each for stereo tracks).

However, if you are just getting started, you are not likely to need many inputs. If you are looking to record only vocals or dialogue, for example, an interface with a single input may get the job done. For the greatest flexibility, look for an audio interface with microphone (XLR) inputs.

If you are operating on a limited budget, you can consider an audio interface that includes just one or two microphone inputs. These may be supplemented with additional line- or instrument-level inputs. Line/instrument inputs can be useful for recording from an electric guitar or keyboard connected directly to the audio interface along with your miked source(s).

Most audio interfaces will include balanced stereo outputs, via a pair of ¼-inch jacks, for connecting to studio monitors. Many also include a stereo headphone jack with volume control. A headphone output is useful for monitoring playback from your DAW while recording: monitoring through headphones instead of your studio speakers helps prevent the playback from bleeding into a live microphone.

Sound Quality

Generally speaking, the quality of a digital audio file is a factor of the sample rate and bit depth used for the recording. But the recorded results are also influenced by the quality of the microphone preamps and the analog-to-digital (or A/D) converters in the audio interface.

Other factors also have a significant impact, including the recording environment and the quality and type of microphone(s) being used to capture the audio.

All of that is to say that you should not consider the specifications of sample rate and bit depth in isolation when seeking to optimize the quality of your recordings.

When to Use High-Resolution Audio

For a simple recording project, you may find yourself using an inexpensive microphone to record audio that has a limited dynamic range (such as when recording dialogue for a podcast or webinar). You may also be recording in an untreated room or office environment. In such cases, recording with a high sample rate and bit depth will not have a beneficial impact: recording at 44.1 kHz and 16 bits will certainly be sufficient.

Other times, you may be recording something with a wider dynamic range, such as a musical performance. This type of recording is often done using higher-quality microphones in a home-studio environment with at least some degree of acoustic treatment. In this case, you might want to consider using higher-resolution audio for better results and more editing flexibility. Recording at 24 bits is recommended in this situation to preserve the quality and dynamic range of the audio being captured.

In cases where you find yourself working in a professional studio environment—with top-of-line microphones, boutique preamps, and other high-end gear—you will almost certainly want to record at a sample rate of 96 kHz or above (and 24 bits) to capture all the subtle nuances of the performance and the character of the associated equipment.

 Recording at 32-bit floating point is typically only necessary when you plan to do extensive additional processing to the audio files after recording. You will hear no audible difference in the quality of the captured audio over recording at 24 bits.

Nearly all of the audio interfaces on the market today support sample rates up to 96 kHz, with many supporting up to 192 kHz. Likewise, 24-bit A/D conversion is a near universal standard for bit depth. This means that your search for an audio interface for Reason Intro will not need to be based on the supported sample rate and bit depth of the device, because almost any device will fit the bill.

 Reason Intro supports audio with sample rates up to 192 kHz.

Instead, look for an audio interface with a construction quality and price that complements your recording goals and budget. For higher-quality audio, look for devices from brand-name companies that have a reputation for high-end gear, but expect to pay a premium for them.

Budget

The ultimate deciding factor often comes down to budget. However, you can weigh the options, such as I/O connectivity and brand reputation, against your budget to find the right compromise.

For example, if you need only a single mike input, you may opt for a high-end audio interface to maximize the sound quality. On the other hand, if having four mike inputs is an absolute requirement, but sound quality is less of a concern, you might select a more utilitarian audio interface instead to stay within budget.

In the end, be sure to consider the budget for your entire recording setup as a whole. You will want to balance the cost of your audio interface against the other considerations covered in this chapter to ensure that you are not spending too much in one area and neglecting others.

Working Without an Audio Interface

In some cases, you may decide to work without an audio interface. You could do this as an interim measure until you can save up for the audio interface you want. You could also do this as a way to run your DAW when you are away from your audio interface, such as when you're on the road with a laptop.

In either case, Reason Intro allows you to utilize the built-in audio on your computer (if available) in lieu of a dedicated audio interface.

On Mac-based systems, no special hardware or software setup is required to work without an audio interface. Launching Reason Intro without an interface connected will cause the software to use **Built-in Microphone + Built-in Output** as the audio device, automatically connecting the available inputs and outputs on your computer to the software.

On some Windows-based systems, Reason Intro will be able to utilize the computer's input and output options right out of the box. Other systems may need a driver installed to allow Reason Intro to access the computer's sound card options. The go-to resource for this is ASIO4ALL, a universal ASIO driver for Windows Driver Model (WDM) audio. ASIO4ALL is an independently developed freeware utility.

 ASIO stands for Audio Stream Input/Output, a computer sound-card driver protocol developed by Steinberg Media Technologies.

In most cases, you simply need to download and install the ASIO4ALL software. Once installed, ASIO4ALL will allow you to run Reason Intro on a Windows machine with no audio interface connected and play back through the computer's built-in speakers or headphone jack.

 To download ASIO4ALL, go to www.asio4all.com and download the latest version in the language of your choice.

Review/Discussion Questions

1.   When a sound event occurs, what are the changes in air pressure called (i.e., the areas where air pressure increases and decreases)? (See "The Basics of Audio" beginning on page 62.)

2.   What is meant by the frequency of an audio waveform? How does the frequency of a sound wave affect our perception of the sound? (See "Frequency" beginning on page 62.)

3.   What characteristic affects our perception of the loudness of a sound? How is loudness measured? (See "Amplitude" beginning on page 63.)

4.   What is the purpose of a traditional microphone? What three basic types of microphones are commonly available? (See "Traditional Microphones" beginning on page 63.)

5.   How are USB microphones different from traditional microphones? What is the advantage of using a USB microphone over a traditional microphone? (See "USB Microphones" beginning on page 64.)

6.   What are the differences between omnidirectional microphones, bidirectional microphones, and unidirectional microphones? Which is the most common type of mike for recording isolated sound sources? (See "Polar Pattern" beginning on page 65.)

7.   What are some accessories available to help you with mike placement, noise reduction, and sound isolation when recording? (See "Microphone Accessories" beginning on page 67.)

8.   How is multitrack recording different from recording to a single audio file? What are some advantages of multitrack recording? (See "What Is Multitrack Recording?" beginning on page 69.)

9.   What two parameters are critical for the analog-to-digital conversion process? What minimum specifications are used for these parameters in modern digital audio systems? (See "The Analog-to-Digital Conversion Process" beginning on page 71.)

10. What is the purpose of an audio interface? (See "The Audio Interface" beginning on page 72.)

11. What are some factors to consider when selecting an audio interface for use with your DAW? (See "Audio Interface Considerations" beginning on page 72.)

12. How is working without an audio interface useful? What additional hardware or software options might you need in order to run Reason Intro without an interface? (See "Working Without an Audio Interface" beginning on page 75.)

 To review additional material from this chapter and prepare for certification, see the Reason Audio Production Basics Study Guide module available through the Elements|ED online learning platform at ElementsED.com.

Selecting Your Audio Production Gear

🎧 Activity

In this exercise, you will define your audio production needs and select components for a home studio based around Reason Intro software. By balancing your wants and needs against a defined budget, you will be able to determine which hardware and software options make sense for you.

🕔 Duration

This exercise should take approximately 10 minutes to complete.

⊕ Goals/Targets

- Identify a budget for your home studio

- Explore microphone options

- Explore audio interface options

- Explore speakers/monitoring options

- Consider other expenses

- Identify appropriate components to complete your system

Getting Started

To get started, you will create a list of needs and define an overall budget for your home studio setup. This budget should be sufficient to cover all aspects of your initial needs for basic audio production work. At the same time, you'll want to be careful to keep your budget realistic so that you can afford the upfront investment. Keep in mind that you can add to your basic setup over time to increase your capabilities.

Use the table below to outline your basic requirements and to serve as a guide when you begin shopping for options. Place an **X** in the appropriate column for your expected needs in each row.

Do not include MIDI gear in this table, as we will address that separately.

 This exercise assumes that you own a compatible computer with built-in speakers for playback. Do not include a host computer in this table.

Function or Component	Not Required	Minimum Configuration	Expanded Configuration
Audio Interface: Input Channels (Recording)	–	1-2 Inputs	4 or More Inputs
Audio Interface: Output Channels (Playback)	–	2 Outputs (for stereo playback)	4 or More Outputs (for stereo playback and output to external gear)
Output Device(s)	Built-In Computer Speakers	Headphones	Stereo Monitor Speakers
Input Device(s) (Microphones)	–	USB Microphone	XLR Microphone(s) (specify type and number)
Accessories and Other (List or Describe) (Stands, Soundproofing, Software Add-Ons, etc.)			

Available Budget for the Above: _____

Identifying Prices

Your next step is to begin identifying prices for equipment that will meet your needs. Using the requirements you identified above as a guide, do some Internet research at an online music retailer of your choice to identify appropriate options for each of the items listed in the table below. You may also want to browse some manufacturers' websites for more information.

Component	Manufacturer and Model	Unit Cost
Audio Interface		
Microphones		
Headphones		
Monitor Speakers		
Accessories/Other		
TOTAL		

Finishing Up

To finalize your purchase decisions, compare the total in the table above to the budget you allocated. If you find that your budget is not sufficient to cover the total cost, you will need to determine which purchase items you can postpone or consider bundle options. On the other hand, if you have money left over in your budget, you can consider upgrade options.

Bundle Options

One way to save some money is to look for gear bundles. These options can be much more affordable than buying the same components separately. Following are some good examples of bundles from Focusrite:

- Scarlett Solo Studio

- Scarlett 2i2 Studio

- iTrack Studio

Each of the above bundles includes a Focusrite audio interface, a large-diaphragm condenser microphone, and a pair of closed-back studio headphones.

For more information, check out the details on the Focusrite website:

- Scarlett interfaces: https://us.focusrite.com/scarlett-range

- iTrack Solo Studio pack: https://focusrite.com/ios-audio-interface/itrack/itrack-solo-studio

Upgrade Options

If you need more audio processing options for Reason, the easiest step would be to upgrade from Reason Intro to the full version of Reason or Reason Suite to gain access to more instruments and effects as well as an unlimited track count.

You can also consider the Rigs Rack Extension bundles available from Reason Studios. More information on these options can be found here:

- Rigs bundles: https://www.reasonstudios.com/reason-rigs

Chapter 4

MIDI Recording Concepts

...What You Need to Record MIDI...

Understanding MIDI is essential to producing music using a DAW. In this chapter, we take a look at some basic MIDI concepts that will get you started creating tracks. We begin by taking a quick look at the history of MIDI and how the MIDI protocol works. Next we check out some of the types of MIDI controllers that are currently available. Then we look at how to set up the controller to communicate with your DAW. This leads into a discussion on the fundamental differences between MIDI and audio data. Finally, we take a quick look at how to begin working with virtual instruments.

✦ Learning Targets for This Chapter

- Understand the basic history of MIDI

- Learn about the MIDI protocol

- Recognize types of MIDI controllers

- Learn how to set up your controller and DAW

- Recognize the differences between MIDI and audio

- Begin tracking with virtual instruments

 Key topics from this chapter are illustrated in the Reason Audio Production Basics Study Guide module available through the Elements|ED online learning platform. Sign up at ElementsED.com.

The term MIDI stands for Musical Instrument Digital Interface. MIDI is a protocol for connecting electronic musical instruments, computers, and other devices, allowing them to communicate with one another.

The MIDI standard was developed in the 1980s to allow musical performance information to be shared between devices. The standard provides specifications for describing musical events, such as note values and durations, allowing musical performances to be represented numerically.

A Brief History of MIDI

Before MIDI, synthesizers and other electronic musical instruments had no standardized way to communicate with each other. Early voltage-controlled analog synthesizers by Bob Moog and Don Buchla (et al.) could send voltage information to each other. But they couldn't communicate more complex musical ideas, such as a discrete note with all of its pertinent information (pitch, amplitude, duration, etc.).

Image courtesy of Perfect Circuit Audio

Figure 4.1 A Buchla 200e analog modular synthesizer

Digital Control

Another major drawback to analog synthesizers was that configurations for individual sounds could not be saved; favorite sounds had to be rebuilt by hand to be reused. Practitioners would actually make little drawings that showed where cables were patched and the position of each knob. (See Figure 4.2.) Some would even take Polaroid pictures so they'd have a photograph of the patch. (Today, this way of working is making a comeback with the current Eurorack synthesizer craze.)

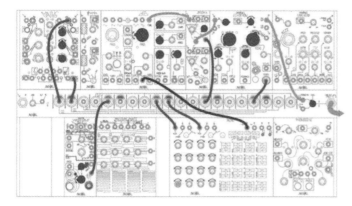

Figure 4.2 A patch "sheet" from a modern Eurorack analog synth

By the early 1980s, despite still using analog circuitry to generate sounds, most synthesizers had made the move to digital control for things like patch storage and recall (saving sounds and recalling them later). This created the possibility of having a common digital communication standard that would allow an electronic musical instrument to speak to any other instrument, regardless of the manufacturer.

The Birth of MIDI

In 1981, a proposal for a "Universal Synthesizer Interface" (USI) was presented at a meeting of the Audio Engineering Society (AES). This led to the original MIDI 1.0 Specification. The specification was finalized in 1983 by a group that would become the MIDI Manufacturers' Association. The term *MIDI* was coined as an acronym for Musical Instrument Digital Interface.

Image courtesy of Perfect Circuit Audio

Figure 4.3 Sequential Circuits Prophet 600 (1982)

At the 1983 National Association of Music Merchants (NAMM) trade show, the first public unveiling of MIDI occurred. A MIDI connection was demonstrated between a Sequential Circuits Prophet 600 (shown in Figure 4.3) and a Roland Jupiter 6. Soon, every major manufacturer had added MIDI connectivity to their synthesizers, drum machines, sequencers, and other devices.

The MIDI standard also helped pave the way for computer-based music. In 1984, Passport Designs released their MIDI/4 sequencer that ran on a personal computer.

The MIDI Protocol

The complete MIDI protocol is quite complicated, but it can be informative to take a look at the basics. MIDI messages can be divided into three basic categories: MIDI notes, program changes, and controller messages.

MIDI Notes

MIDI notes are the most prevalent part of the protocol. Note data is the building block for almost all the work you'll do with MIDI.

Note Values

One thing to understand about MIDI messages is that almost all parameters have a range of values spanning from 0 to 127. So, for instance, the pitch of a MIDI note is represented using a range of 0 to 127. A value of 0 represents the C note two octaves below the bottom of a piano. We refer to this note as C-2.

Middle C is typically assigned a note value of 60 and referred to as C3 (although some manufacturers use C4 instead). Figure 4.4 provides a diagram of note values.

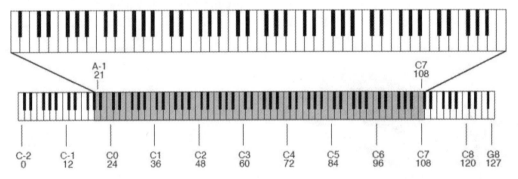

Figure 4.4 A standard piano keyboard with note names and MIDI note numbers

Note Velocity

The other important bit of note information is the velocity, or how hard a note is played. Velocity is also represented using a range of 0 to 127, with the value of 64 falling right in the middle (sort of a *mezzo forte* in musical terms). Playing a note lighter will result in a lower velocity value, whereas playing harder will result in a higher value.

Note On/Off Events

A MIDI note is a combination of a note number (pitch) and a velocity number. We refer to this combination as a **NOTE ON** command, because it activates the note in a technical sense. The **NOTE ON** is eventually followed by a **NOTE OFF** command, which tells the device to stop playing the note. The **NOTE OFF** command is the same note number, paired with a velocity of zero. The Note On and Note Off events are really all we need to send a note between two MIDI devices.

Program Changes

The ability to store and recall patches was an early advantage of digital control for synthesizers. The MIDI protocol improved upon this, enabling patch change commands (also known as program changes) to be sent to and from devices.

Like other MIDI messages, patch change commands use a range of values from 0 to **127**. Thus, most banks of sounds on a MIDI keyboard will have 128 sounds. Table 4.1 shows an example bank of 128 sounds that many synthesizers support. This is part of the General MIDI standard.

Table 4.1 General MIDI Instrument Families

Program Change #	Family Name	Program Change #	Family Name
1 – 8	Piano	65 – 72	Reed
9 – 16	Chromatic Percussion	73 – 80	Pipe
17 – 24	Organ	81 – 88	Synth Lead
25 – 32	Guitar	89 – 96	Synth Pad
33 – 40	Bass	97 – 104	Synth Effects
41 – 48	Strings	105 – 112	Ethnic
49 – 56	Ensemble	113 – 120	Percussive
57 – 64	Brass	121 – 128	Sound Effects

Controller Messages

The final category of MIDI data is that of controller messages—also known as MIDI continuous control (CC) messages. These are messages that can be used for purposes such as capturing the expressiveness of a MIDI performance (pitch bend or mod wheel movements) or performing studio mixing tasks (volume or pan changes). MIDI CC data represents an essential part of how keyboardists actually *play* their instrument.

Table 4.2 provides a list of standard MIDI continuous control defaults. Notice that certain controller numbers have their functions designated as "undefined," leaving room to make custom assignments.

Table 4.2 MIDI Controller Numbers (Excerpt)

Controller Number	Hex	Controller Name/Function
0	00h	Bank Select
1	01h	Mod Wheel
2	02h	Breath Controller
3	03h	Undefined
4	04h	Foot Controller
5	05h	Portamento Time
6	06h	Data Entry MSB
7	07h	Main Volume
8	08h	Balance
9	09h	Undefined
10	0Ah	Pan

MIDI Controllers

To get started working with MIDI, it helps to have a MIDI controller. A MIDI controller is a device that generates MIDI performance data and transmits it to another device, such as a DAW, for processing. You can use a MIDI controller for recording MIDI notes and performance data into your DAW or for triggering a sound module or virtual instrument.

A variety of different MIDI controllers are available today. The most common designs continue to be based on the piano. With origins in the 18th century, the familiar black and white piano keyboard is a staple in the MIDI controller market. However, the last decade or so has seen an explosion in MIDI controller innovation. Today's options include drum pad controllers (inspired by the drum machine designs of the 1980s), grid controllers (originally introduced to augment the grid-based interface of Ableton Live), and mixer controllers (borrowing their faders and knobs from established analog and digital mixer designs).

Tone-Generating Keyboards

Many keyboard-based devices on the market can be classified as *tone-generating keyboards*. These are keyboards capable of generating their own sounds. Although generally designed for stand-alone performance, most tone-generating keyboards also provide MIDI connectivity. This allows you to use the keyboard to generate MIDI performance data and record it on your DAW. It also allows the keyboard to receive MIDI data from elsewhere and use it to trigger the device's onboard sounds.

Tone-generating keyboards generally fall into two categories—digital pianos and synthesizers:

- **Digital Pianos**—Digital pianos are exactly what the name implies: they look like scaled-down versions of a real piano. Digital pianos typically feature a full 88-key keyboard (often with simulated piano action known as *hammer action*). This type of keyboard is designed to compete head-to-head with real acoustic pianos; thus, they are very simple to operate.

  Digital pianos generally offer a limited range of sounds (such as piano and organ). However, more expensive models may include additional sounds and auto-accompaniment features that can generate bass lines and drum patterns. Perhaps the most distinguishing feature of a digital piano is the presence of built-in speakers, which synthesizers rarely offer. (See below.)

Figure 4.5 Yamaha Arius Digital Piano

- **Synthesizers**—The term *synthesizer* encompasses a wide range of instruments. The term was originally used to describe analog devices that employed a combination of oscillators and filters to create sounds. These devices generated approximations of acoustic instruments (such as flute or violin) as well as otherworldly sounds that didn't bear resemblance to traditional instruments.

  But synthesizers quickly evolved to encompass both analog and digital technologies, with newer incarnations capable of using older analog-style synthesis as well as modern digital techniques such as sampling, physical modeling, and frequency modulation (FM) synthesis.

  Modern synthesizers come in a variety of sizes, ranging from small, highly portable two-octave (25-key) models to full 88-key piano-sized models.

Image courtesy of Perfect Circuit Audio

Figure 4.6 Moog's classic Minimoog synthesizer (1970)

If you already own a digital piano or synthesizer, look for a MIDI 5-pin connector or USB port for connecting it to your DAW. If you don't already own such a device, you can consider using a keyboard controller instead.

Keyboard Controllers

Keyboard controllers are essentially piano-style controllers that provide MIDI output. You can use these to create or record a MIDI performance. However, they do not include any onboard sound or tone-generating capabilities. This means that you cannot use a MIDI controller as a stand-alone instrument. Instead, the keyboard is designed to control other hardware synthesizers, sound modules, or virtual instruments.

The keyboard controller serves as a kind of universal MIDI input device. Using a piano keyboard as the performance interface, keyboard controllers enable any user with basic piano skills to create almost any kind of MIDI performance with little to no added learning curve.

Keyboard controllers come in a variety of sizes and prices. Users on a limited budget can consider a basic 25- to 37-key controller for less than a hundred dollars. More sophisticated models with up to 88 keys can be found in the $150 to $500 range. These commonly include knobs, faders, and drum pads that can be mapped to control almost any function of the target instrument.

Figure 4.7 M-Audio Oxygen 49 keyboard controller

Drum Pad Controllers

Most modern drum pad controllers are descended from Roger Linn's ingenious designs from the 1980s. The most iconic of Linn's designs was the Akai Professional MPC60, introduced in 1988. The 4 × 4 layout of pads that Linn pioneered has since become a standard pad arrangement for dozens of products.

The pads themselves are carefully designed to have a great feel that helps the player to perform dynamic drum and percussion grooves.

Figure 4.8 Akai Professional MPD218 drum pad controller

Grid Controllers

Today, the grid controller is probably the most popular controller type after the piano-style keyboard. Grid controllers have become popular for studio work, but they have also been embraced by live performers in a variety of genres, including EDM and hip-hop.

The grid controller design was inspired by Ableton Live's Session view, with each pad used to trigger clips, play drum sounds, and adjust controls such as volume. Some grid controllers, such as Novation's Launchpad Pro and Ableton's Push 1 and 2, can be used to enter notes as well. These include options for restricting notes to a musical scale so that no wrong notes can be played! While grid controllers are generally optimized to work with Ableton Live, they can be used with an increasing number of DAWs.

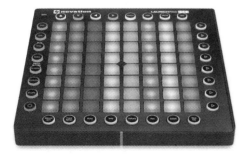

Figure 4.9 Novation Launchpad Pro grid controller

Alternate Controllers

You can also find some very innovative new controller styles on the market these days, offering alternatives to the standard keyboard, pad controller, or grid controller.

A primary focus of such alternative controllers currently is MIDI Polyphonic Expression (MPE). MPE is a new industry-wide standard that uses a separate MIDI channel for each touch. This channel-per-note configuration permits the controller to transmit discrete vibrato, glissando (pitch slides), note pressure, and other expressive information for each note that is played.

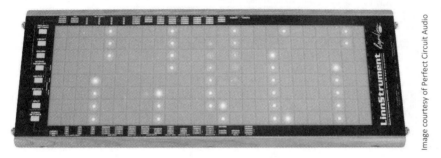

Image courtesy of Perfect Circuit Audio

Figure 4.10 Roger Linn's LinnStrument is one example of an alternate controller.

What to Look for in a MIDI Controller

Deciding which controller or controllers to purchase can be an overwhelming task! The most important thing is to determine which activities you will be performing frequently, which you'll be doing only occasionally, and which you won't be doing at all. Here are some suggestions based on the strength of various controller types for typical music production tasks:

- Entering notes:
  - Keyboard (best choice)
  - Grid (good alternative)
- Playing beats:
  - Drum pad (best choice)
  - Grid (good alternative)
  - Keyboard (also good)
- Arranging:
  - Grid

Purchasing a Keyboard Controller

If you decide to purchase a keyboard controller, you'll need to decide on a few other attributes. The most important is probably the number of keys (note range). Your needs here will generally be determined by your proficiency as a keyboard player.

If you are an accomplished keyboardist who plays with two hands simultaneously, you will probably feel most comfortable with a keyboard featuring 61 keys or more. A full-sized, 88-key controller will offer the widest note range and the best feel. On the other hand, if you play using just one hand at a time, you may be satisfied with a smaller 25-key model.

Aside from note range, you'll also want to consider a device's support for velocity sensitivity, aftertouch, pitch and mod wheel controls, and other mappable controls like knobs and sliders.

Velocity Sensitivity

As mentioned previously, capturing the intensity of each note is often a key ingredient to the expressiveness of a performance. Most moderately sophisticated keyboard controllers will be velocity sensitive, meaning they measure how hard each key is struck. However, some budget models may not include this capability, representing a significant limitation for many types of MIDI recording.

Aftertouch

Aftertouch lets a performer add expressive inflections to a sustained note or chord. Aftertouch can be used to add vibrato, pitch bends, and swells by varying the pressure on the held keys. When using a virtual instrument that supports it, aftertouch can be essential to making the instrument sound realistic. (Think of guitar solos and horn parts, for example: how often are sustained notes completely static?) Here again, budget keyboard controllers often do not include support for this parameter. Be prepared to do some comparison shopping when trying to get the most bang for your buck.

Pitch and Mod Wheel Controls

These are additional options for adding expressiveness to a performance. Pitch bend and mod wheel controls are common on synthesizers but may not be included on some MIDI controllers. A traditional pitch wheel adds pitch bend (portamento) control, allowing the player to bend a note (or chord) up or down in a continuously variable manner. Pitch bend controls are sometimes included in non-wheel format, such as in a joystick, knob, or ribbon format.

Modulation wheels (or mod wheels) are typically used to set a vibrato amount. However, the mod wheel can also be mapped to other parameters, such as volume swells, tremolo, and filter sweeps.

Other Mappable Controls

Additional buttons, knobs, and sliders may be included on midsized and full-sized keyboard controllers. These controls can be mapped to plug-in parameters and other MIDI controls. In some cases, they can even operate functions in your DAW, such as transport controls and volume faders.

Setup and Signal Flow

Once you've selected a MIDI controller, you'll need to establish communication between the controller and your computer. Almost all modern MIDI controllers use USB rather than the traditional MIDI cable to communicate. Configuring such a device can be as simple as plug and play, if the device is USB class-compliant. However, some devices require that you install specific drivers built for your computer and operating system.

Plug-and-Play Setup

For class-compliant USB devices, no drivers are required to communicate with your computer. You can verify that a device is class-compliant by checking the specifications on the manufacturer's website. Alternatively, you can connect the device to your computer and launch the appropriate software utility to verify communication.

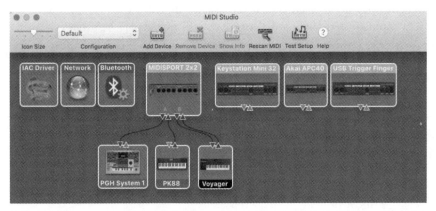

Figure 4.11 An example MIDI setup in Apple's Audio MIDI Setup utility; class-compliant devices will appear here automatically.

MIDI Cables and Jacks

If you are using a device that has regular MIDI connections (not USB), you will have to deal with MIDI cables and jacks. A MIDI cable is technically called a 5-pin DIN cable and has connectors as shown in Figure 4.12.

Figure 4.12 A pair of MIDI cables

MIDI data is actually transmitted using only two pins on the 5-pin connector.

An important aspect of a MIDI cable is that it can carry *16 channels* of information. That means you could have up to 16 patches playing on one synthesizer, and they could all be addressed individually using a single MIDI cable.

The jacks that MIDI cables plug into are typically called *ports*. These can include an input, an output, and sometimes a "thru" connection.

Figure 4.13 A typical configuration of MIDI ports on a device

The MIDI In port receives incoming MIDI data from an outside source; this is commonly used to play back a MIDI performance coming from a DAW or other source using the onboard sounds of the device. The MIDI Out port sends MIDI data out from the device, for recording to a DAW or for triggering a different sound module. The MIDI Thru port passes the signal from the In port directly out to another device, without modifying it by any performance being done on the device. This can be very handy if you have a small MIDI interface with only a couple of ports, or if you have a complex live performance setup and don't want to lug an interface around.

Using a MIDI interface

Although most modern controllers use USB, you may encounter hardware devices (such as older synthesizers) that use traditional 5-pin MIDI cables to communicate. In these cases, you'll need a way to connect the MIDI cables to your computer, because the computer will not have the appropriate jacks for MIDI. You can accommodate these connections using an audio interface with MIDI jacks (quite common on all but the smallest interfaces).

Figure 4.14 MIDI jacks on the back of a Focusrite 2i4 audio interface

As an alternative to an audio interface, you can use a dedicated MIDI interface to route MIDI data to your computer. MIDI interfaces come in a variety of shapes and sizes and can have as many as eight jacks for MIDI input and output.

Figure 4.15 M-Audio Uno MIDI interface

Considerations for Using Multiple MIDI Devices

Even the smallest studio will often have more than one MIDI device. This could involve any combination of keyboard, drum pad, grid, and mixer controllers. Aside from the ergonomic considerations of placing multiple controllers in the physical space, you'll also need to decide whether you want the devices to control specific instruments or to all control the currently selected instrument.

In most DAWs you can leave inputs on your MIDI tracks assigned to ALL so that the currently selected instrument will respond to any controller you touch. This makes it easy to play a keyboard for a piano part, then switch to drum pads to perform a beat, without having to stop and change the MIDI routing.

Reason uses a system called Master Keyboard Input to route MIDI input to devices and virtual instruments. One MIDI keyboard controller is designated as the Master Keyboard. The device that responds to the Master Keyboard can be easily changed in the Reason interface. By default, all other controllers also follow along with the Master Keyboard, making it easy to switch control devices without changing MIDI routing.

MIDI Versus Audio

One of the more difficult concepts to grasp when you begin working with a DAW is the difference between MIDI and audio. While most novice audio producers have a decent understanding of how audio data is recorded and represented in a DAW, many do not have a similar understanding with regard to MIDI data.

What Is MIDI?

As discussed earlier in this lesson, MIDI is a protocol that facilitates communication between a huge range of hardware and software devices. MIDI data is not audio, but rather a series of messages that can communicate information such as notes (pitch), duration, velocity (how hard a key or pad is pressed), dynamics (volume), and more. A tone-generating hardware or software instrument can receive MIDI messages and convert them into audio data.

In other words, the MIDI data has the potential to become music that we can hear, but there's no way to listen to the MIDI data by itself.

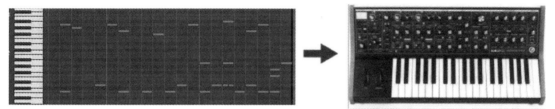

Figure 4.16 MIDI data (in piano roll format) must be sent to a tone-generating device to become music that we can hear.

MIDI messages share some features with traditional music notation. Both contain the information necessary to play a piece of music, but neither is capable of actually becoming music on its own. Like MIDI, the notes of a musical score aren't audio. Instead, they represent the potential for audio once they are played using an instrument.

MIDI and notation contain information that can convey essential musical elements such as pitch, duration, velocity, and dynamics. Many DAWs allow MIDI data to be viewed and edited as music notation (in addition to piano roll and other more modern formats). Reason Intro provides only piano roll editing.

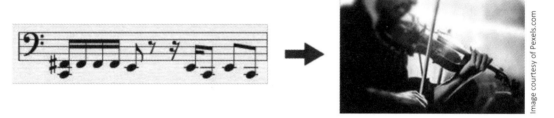

Figure 4.17 Music notation must be read by a musician to become music we can hear.

Monitoring with Onboard Sound Versus Virtual Instruments

If you are using a tone-generating keyboard such as a digital piano or synthesizer, you will need to decide whether you want to monitor from the onboard sound of the device or, alternatively, silence the onboard sounds and instead listen to a virtual instrument inside of your DAW.

Performers rarely want to hear the sound of the controller while recording MIDI data into a DAW. (This can cause an annoying doubling of the controller audio with the DAW audio.) Typically, you'll want to monitor from the associated virtual instrument's output alone. This can be accomplished by turning off Local Control on the keyboard device. This setting will disable the internal sounds of the device while still sending MIDI data out to the DAW.

You may need to refer to the user guide or manufacturer's website to find the local control setting on your particular device. Alternatively, you can usually simply turn down the audio output of the controller and listen only to the DAW.

Tracking with Virtual Instruments

Most modern music production studios incorporate software instruments, or virtual instruments, into their toolkit. In fact, a number of software manufacturers are known primarily for their virtual instrument offerings, including Native Instruments, Arturia, and Spectrasonics. Virtual instruments encompass a huge range of musical devices, including digital synthesizers, analog-modeling synthesizers, samplers, drum machines, and much more. They have become so ubiquitous in the DAW world that every major manufacturer offers a range of free or paid virtual instruments.

Figure 4.18 Native Instruments Kontakt sampler

Creating Tracks for Virtual Instruments

Working with virtual instruments in your DAW can seem a bit confusing at first. It's not quite as simple as just creating a track and pressing record! The first step to working with virtual instruments is to create the appropriate track type, which is usually referred to as an Instrument or MIDI track, depending on the DAW. These tracks are capable of recording and editing MIDI data, routing the MIDI data to a virtual instrument, and then mixing the virtual instrument's resulting audio output.

Let's take a look at a couple of example configurations.

Virtual Instruments in Reason Intro

Routing MIDI to a virtual instrument in Reason Intro is similar to the process in many other DAWs, including Pro Tools, Cubase, and Logic.

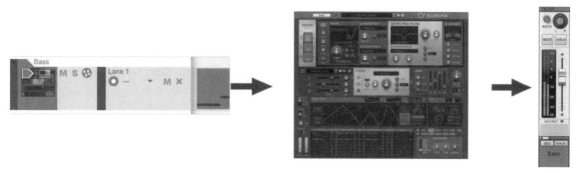

Figure 4.19 When MIDI note data is routed to the Europa instrument, the device's audio output is processed in the associated mixer channel.

In Figure 4.19, the **Bass** track sends its MIDI data to the Europa virtual instrument device, which is routed to a channel in the mixer. Then, Europa converts the MIDI data to audio, with the sound being determined by the patch/preset that is currently selected. Next, the audio output from Europa is routed through the Bass mixer channel, where its volume, pan, and other audio attributes can be adjusted as desired. Finally, the audio output of the mixer is routed to the audio interface, so that it can be monitored through speakers or headphones.

Virtual Instruments in GarageBand

GarageBand provides another example of this concept, although the approach used in the software is somewhat different. (See Figure 4.20.) Note that GarageBand doesn't offer a Mixer view, so the track is only visible in a horizontal layout that is similar to the Sequencer view in Reason Intro.

In GarageBand, the sound module is assigned by selecting the desired source from the Sound Library, rather than by placing a virtual instrument plug-in on the track.

Figure 4.20 Using a drum instrument in GarageBand

In Figure 4.20, we can see the MIDI data in the familiar piano roll format on the track. That MIDI data is being routed to the East Bay drum kit (whose controls are displayed at the bottom). The kit converts the MIDI data to audio. The audio output actually appears on the horizontal controls just to the left of the MIDI data, where the volume and pan attributes can be modified. From there, the audio goes to the interface output.

Summary

The function of a virtual instrument is quite similar regardless of the DAW you choose. Professional-level DAWs offer sophisticated routing and additional views for working with MIDI data (as well as audio). By contrast, entry-level DAW options (such as GarageBand) aim to keep things as simple as possible. Regardless of where you begin, once you have gained basic familiarity with your chosen DAW, you'll find it relatively easy to apply that knowledge to any other DAW you should encounter.

Review/Discussion Questions

1. What were some of the drawbacks of historical analog synthesizers? (See "A Brief History of MIDI" starting on page 84.)

2. How did digital control of synthesizers pave the way for the development of MIDI? (See "Digital Control" starting on page 84.)

3. What are the three basic categories of MIDI messages? (See "The MIDI Protocol" starting on page 86.)

4. What is the standard range of values for all MIDI messages? (See "Note Values" starting on page 86.)

5. What MIDI note number is typically used to represent middle C on a piano? Is this number the same on every device? (See "Note Values" starting on page 86.)

6. What are the two components of a MIDI Note On command? (See "Note On/Off Events" starting on page 86.)

7. What is another name for a patch change command when using MIDI? (See "Program Changes" starting on page 87.)

8. What type of MIDI data can be used to increase the expressiveness of a MIDI performance? What are some examples of these types of MIDI messages? (See "Controller Messages" starting on page 87.)

9. Aside from keyboards, what are some other types of MIDI controllers? (See "MIDI Controllers" starting on page 88.)

10. What are two categories of tone-generating keyboards? (See "Tone-Generating Keyboards" starting on page 88.)

11. What does it mean when a USB device is labeled "class-compliant"? What is the advantage of a class-compliant USB device? (See "Plug-and-Play Setup" starting on page 94.)

12. How can MIDI notes be converted to an audio signal that we can hear? (See "MIDI Versus Audio" starting on page 96.)

13. Describe the basic signal flow used with virtual instrument devices in Reason Intro. (See "Tracking with Virtual Instruments" starting on page 98.)

Selecting Your MIDI Production Gear

🎧 Activity

In this exercise, you will define your MIDI production needs and select components to complement the Reason system you configured in Exercise 3. By identifying the type of MIDI production work you will be doing, you will be able to select an appropriate MIDI controller or controllers to meet your needs.

🕐 Duration

This exercise should take approximately 10 minutes to complete.

⊕ Goals/Targets

- Identify a budget for your MIDI hardware

- Explore controller options

- Consider other expenses

- Identify appropriate components to complete your system

Getting Started

To get started, you will create a list of requirements for your MIDI controller and define an overall budget for your MIDI setup. Once again, make sure your budget is sufficient to cover all your immediate needs, while remaining realistic about what you can afford. You can always add to your basic MIDI setup over time with more devices and virtual instruments.

Use the table below to identify your basic MIDI requirements and to serve as a guide when you begin shopping for a controller. Place an **X** in the appropriate column(s) for each row.

Criteria	Rating			
Piano Keyboard Skills	Novice / Hunt and Peck	Competent Beginner	Intermediate	Advanced
Intended Uses	Drum Programming, Effects, Short Musical Parts, Loops, EDM	Background Music Production, Simple Keyboard Parts and Other Instruments	Lead Instrument Parts, Two-Handed Performances, Complex Arrangements	Orchestral Scores and Arrangements
Feature Requirements	Basic MIDI Input	Velocity Sensitivity	Pitch Bend and Mod Wheel	Aftertouch
Accessories and Other (List or Describe) (Keyboard Stands, Foot Pedals, etc.)				

If your **X**'s fall predominantly in the left half of this table, you may want to consider a drum pad or grid controller or look for a budget-model keyboard controller. If your **X**'s fall predominantly in the right half, you should probably target a full-sized and fully featured keyboard controller. If your MIDI needs are extensive, you may want to consider selecting more than one type of controller to purchase.

Available Budget for MIDI Gear: _____

Identifying Prices

Your next step is to begin identifying prices for the type of controller and accessories that will meet your needs. Using the requirements you identified above as a guide, conduct some research online to identify available options. List potential matches in the table below, along with short descriptions (including number of keys, if applicable) and prices.

Type of Controller	Manufacturer and Model/Description	Price
Accessories/Other		

Total Cost for MIDI Gear: _____

Finishing Up

To finalize your purchase decisions, total the cost of each option plus your required accessories and compare them to the budget you allocated. If you find that your budget is not sufficient to cover your preferred choice, consider a different option and plan to upgrade in the future.

If you are able to purchase your first-choice items and have money left over, you can look at software add-ons to supplement your collection of virtual instruments. For example, you might consider upgrading to the full version of Reason or Reason Suite, which will add a number of additional instruments and effects.

You can also consider the Rigs Rack Extension bundles available from Reason Studios. More information on these options can be found here: https://www.reasonstudios.com/reason-rigs.

Chapter 5

Reason Concepts, Part 1

...What You Need to Know to Get Started with Reason Software...

This chapter introduces you to some basic operations and functions you need to be familiar with to get started working in Reason software. We cover how to create a new project, how to access and use the main views, and how to create tracks. We also cover basic navigation and selection techniques, including transport controls and zooming and scrolling operations. These foundational concepts will help you get up and running and remain productive as you work.

⊕ Learning Targets for This Chapter

- Learn how to create a new Reason Intro song

- Recognize uses for the Sequencer view, Main Mixer view, and Racks view

- Become familiar with the basic controls in each main view

- Learn how to create tracks and understand the supported track types in Reason

- Learn how to navigate in the Sequencer view by scrolling and zooming

- Learn playback and selection techniques

Key topics from this chapter are illustrated in the Reason Audio Production Basics Study Guide module available through the Elements|ED online learning platform. Sign up at ElementsED.com.

If you need professional-level audio production capabilities on a limited budget, Reason Intro is an excellent place to start. The techniques and workflows you use in Reason Intro will transfer seamlessly to standard Reason software (and Reason Suite), should you decide to upgrade in the future.

Creating a Song

When you launch a Reason product (Reason Intro, Reason, or Reason Suite), the first thing you'll see after starting up and logging in is an empty song document, which is ready for you to start creating music. If desired, you can use Reason's preferences to customize the startup behavior, so that it opens a template you've created or the last song you were working on.

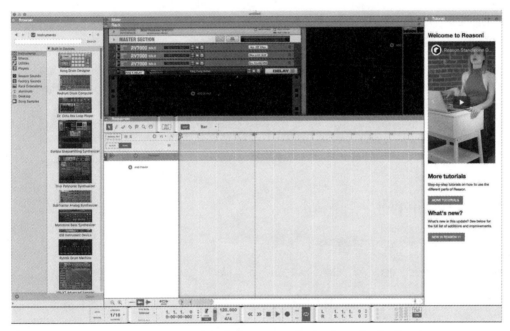

Figure 5.1 An empty song document created after starting Reason Intro for the first time

If Reason Intro is already running, you can also create a new song by choosing FILE > NEW.

 Reason allows you to have multiple songs open at once, which can be useful for copying content between files, for example. However, leaving song files open can greatly increase the memory and CPU usage of Reason. Be sure to close any files you aren't using. In general, try to keep only one song open at a time.

Main Views and Panels

The main views in Reason include the Sequencer view, the Main Mixer view, and the Racks view. All of these views share the screen with the Browser on the left and the Transport Panel at the bottom.

Each of the main views can be accessed under the **WINDOW** menu at the top of the screen. The Browser panel and Transport Panel can also be accessed under the **WINDOW** menu.

The sections below cover the features of each of these main views and panels in Reason Intro.

Sequencer View

The Sequencer view is where you will do most of your work recording, arranging, and editing audio and MIDI data. This window provides a timeline display of the audio waveforms and MIDI performances on tracks in your song.

To access the Sequencer view when it is not displayed, choose **WINDOW > VIEW SEQUENCER**.

 You can also display the Sequencer view by pressing function key F7.

The audio and MIDI data on your tracks will be displayed as *clips* in the Sequencer view. Clips are representations of audio or MIDI content in a Reason song.

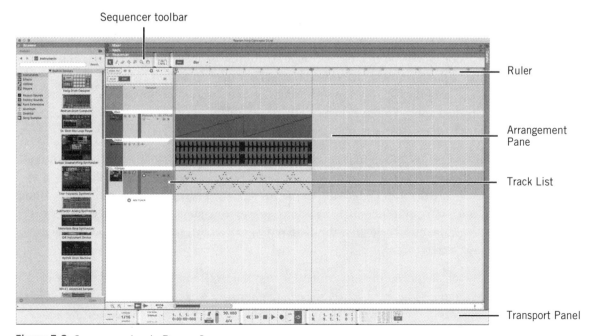

Figure 5.2 Sequencer view in Reason Intro

Sequencer Toolbar

The toolbar area at the top of the Sequencer provides a variety of functions, including the sequencer tool buttons, the Edit Mode button(s), the grid settings, and the Inspector.

Sequencer Tool Buttons

The buttons on the left side of the toolbar activate tools that determine how the mouse functions within the sequencer.

Figure 5.3 Sequencer tool buttons (Selection tool active)

- **Selection Tool**—This tool allows you to select, move, and resize clips on tracks within the sequencer. You can click on a clip to select it or you can click and drag to move it. If you click on an area of the sequencer that does not contain any clips and then drag, you can draw a marquee selection rectangle to select multiple clips. When a clip is open in an edit mode, the Selection Tool is also used to select and move events within the clip, such as MIDI notes. The Selection Tool is the default tool and is the primary arranging tool in Reason Intro. You will use this tool most of the time.

- **Pencil Tool**—This tool allows you to modify content within the sequencer. If you click with the Pencil Tool in an empty area of the sequencer, you can create new, empty clips. When a clip is open in an edit mode, the Pencil Tool can change the contents of the clip. For example, when editing a note clip (aka a MIDI clip), you can use the Pencil Tool to draw in notes and modify note velocities. The Pencil Tool also has an alternate mode (**Draw Multiple Notes**) that allows you to draw a series of individual notes by clicking and dragging when editing a note clip.

- **Eraser Tool**—This tool allows you to remove content within the sequencer. If you click a clip with the Eraser Tool active, the clip will be deleted. When a clip is open in an edit mode, the Eraser Tool can remove the contents of the clip. For example, when editing a note clip, you can use the Eraser Tool to delete individual notes. You can click and drag with the Eraser Tool to draw a marquee selection rectangle that will erase everything within the rectangle.

- **Razor Tool**—This tool allows you to split content within the sequencer. If you click a clip with the Razor Tool active, the clip will be split into two clips at the clicked location. When a clip is open in an edit mode, you can use the Razor Tool to split the contents of the clip. For example, when editing a note clip, the Razor Tool can split a single note into two separate notes. The Razor Tool follows the grid snapping settings (discussed below); if the Snap setting is enabled, the split location will align with the nearest grid increment.

- **Mute Tool**—This tool allows you to mute content within the sequencer so that it doesn't play back. If you click a clip with the Mute Tool active, the clip's appearance will change to indicate that it is muted, and the contents of the clip will no longer be audible during playback. When editing a note clip, you can use the Mute Tool to mute individual notes within the clip.

- **Magnifying Glass Tool**—This tool allows you to zoom horizontally and vertically in the sequencer. If you click in the sequencer with the Magnifying Glass Tool active, the sequencer will zoom in, centering on the point where you clicked. You can click multiple times to continue zooming. You can click and drag with the Magnifying Glass Tool to draw a marquee selection rectangle around content in the sequencer; the sequencer will zoom in on the content in the rectangle when you release the mouse.

 Hold the Ctrl key (Windows) or the Option key (Mac) to change from zooming in to zooming out when using the Magnifying Glass Tool.

- **Hand Tool**—This tool allows you to scroll the arrangement area of the sequencer. When the Hand Tool is active, you can click with the mouse and drag in any direction to move the arrangement in that direction.

- **Speaker Tool**—This tool allows you to audition or preview certain areas of audio information within the sequencer. The Speaker Tool is displayed contextually, so you may not see it within the toolbar. It will appear when you open an audio clip in one of the audio edit modes (Slice Edit, Pitch Edit, or Comp Edit, covered below).

Edit Modes

The Edit Mode buttons allow you to open the clips in your tracks for editing. The set of buttons that appears changes contextually depending on the type of track or clip you select in the sequencer. When you have an audio clip selected, the buttons shown in Figure 5.4 appear.

Figure 5.4 Audio edit modes in Reason Intro

You don't necessarily need to use any of these edit modes when you are starting out with Reason. You can perform basic audio editing without any of these specialized tools. You can arrange, trim, stretch, cut, copy, paste, and split audio all directly within the Arrangement Pane without activating an audio edit mode. However, it can be helpful to be aware of the edit modes and have a basic understanding of their functions.

- **Slice Edit**—This mode analyzes an audio clip and identifies its transients, or peaks in the waveform. Reason adds Slice Markers at each transient. These allow you to warp or stretch pieces of audio, split the audio at the Slice Markers, or convert the audio to a loop that can be used with Reason's loop players.

- **Pitch Edit**—Reason's Pitch Edit Mode provides similar functionality to Antares Auto-Tune or Apple Logic's Flex Pitch, allowing you to adjust the pitch of recorded audio. This mode is designed for use on "monophonic" audio where only a single note is heard at a time, like a vocal recording of one singer.

■ **Comp Edit**—Comping (short for compositing) is a technique that involves recording the same part numerous times and stitching together the best parts of each recorded take to create a polished final product. When you record over the same area multiple times in Reason, you build up a clip that contains these multiple recorded takes. The Comp Edit Mode allows you to view the various included takes and assemble a composite version of the performance.

When you do not have an audio clip selected, you will see only a single edit mode button, simply labeled Edit Mode.

Figure 5.5 Edit Mode button in Reason Intro

With a MIDI note clip selected on an instrument track, you can click this button to open the clip for editing. While Edit Mode is active, you can manipulate the clip's contents by adding notes, deleting notes, changing note lengths and velocities, and so on.

 You can also open a note clip for editing by double-clicking the clip with the Selection Tool or by pressing Enter/Return when the clip is selected.

Grid Settings

The grid controls are used to change the way you are able to move and edit content in the sequencer, relative to the musical grid of bars and beats.

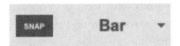

Figure 5.6 Grid settings in Reason Intro

The Snap button determines whether objects like clips and notes lock to the grid as you move and edit them. It also affects how tools like the Razor Tool work. When Snap is turned off, you can move and edit content freely; when Snap is turned on, Reason will constrain content to the grid as you move and edit material.

The musical increment that determines the size of the grid is selected in the drop-down menu next to the Snap button. Changing the grid size determines how the Snap mode functions. The displayed grid lines will also visually change in the sequencer.

If you choose a very small grid size (e.g., 1/64 notes), you may need to zoom in to see all of the grid lines in the sequencer.

 Many DAWs include grid editing modes with grid increments based on alternative timescales, such as minutes:seconds or timecode. As a music composition tool, Reason restricts the grid to musical note increments.

Inspector

The Inspector area in the toolbar allows you to view and change the numeric values that control a clip in the sequencer, like the exact musical position and length of a clip. The Inspector is displayed contextually, so you won't see it at all if you don't have a clip or an event selected. Similarly, the fields displayed in the Inspector change depending on the type of content you have selected, such as an audio clip or a note clip.

Position	Length
5. 1. 1. 0 ⬍	4. 0. 0. 0 ⬍

Figure 5.7 Note clip Inspector in Reason Intro

Position	Length	Fade In	Fade Out	Level (dB)	Transpose
1. 1. 1. 0 ⬍	8. 0. 0. 0 ⬍	0. 0 ⬍	0. 0 ⬍	0.00 ⬍	0.00 ⬍

Figure 5.8 Audio clip Inspector in Reason Intro

The Ruler

The Ruler appears at the top of the Sequencer view, below the Sequencer toolbar and above the tracks and Arrangement Pane area. The Ruler provides measurement indicators that help you identify specific locations on a song's timeline. The Reason Ruler shows the song's timeline in musical bars and beats. The musical subdivisions shown in the Ruler will change based on the grid size you choose and your zoom level in the sequencer.

 Many DAWs include multiple rulers that can each display different timescales, such as minutes:seconds or timecode. As a music composition tool, Reason provides a ruler showing musical note increments only.

Track List

The Sequencer view in Reason includes a Track List on the left side of the arrangement area. The Track List consists of the track headers next to the track lanes that contain the clips for the tracks in the arrangement area. The Track List provides controls for each track for functions such as muting and soloing tracks, enabling or disabling tracks for recording, collapsing or expanding tracks, selecting tracks, changing the order of the tracks in the sequencer, and more.

Browser Panel

The Browser Panel provides access to all of the instruments, sounds, devices, and plug-ins that you will use in your Reason productions. Using the Browser, you can create devices from the four main categories used by Reason: Instruments, Effects, Utilities, and Players.

 To toggle the display of the Browser Panel, choose Window > Show/Hide Browser or press Function Key F3.

When working in the Sequencer view, you'll find the Instruments category is the most useful. You can add new instruments to your song by dragging the icon for an instrument from the Browser to the bottom empty area of the Track List. You can also drag to a location between two existing tracks. Either action will automatically add the instrument to the Racks View, connect it to a Mix Channel in the Main Mixer view, and add a new track for the instrument in the Sequencer.

Similarly, you can drag an instrument from the browser on top of an existing instrument track to replace the instrument currently assigned to that track.

The Browser also acts like a Finder window (for Mac systems) or a File Explorer window (for Windows systems). You can use it to navigate the files and folders on your computer to locate audio files that you want to add to your song. You can then drag the audio files into the sequencer to create new audio tracks or to add audio to existing tracks.

Transport Panel

The Transport Panel at the bottom of the window provides buttons for various transport functions, such as play, stop, fast-forward, and rewind. This panel also displays Position/Time placements, Loop Locator placements, system status indicators, and more.

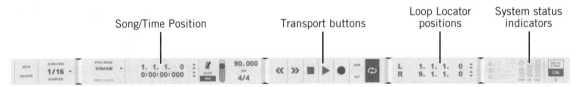

Figure 5.9 Transport Panel in Reason Intro

Song/Time Position indicators. The Song Position and Time Position area of the Transport Panel indicates the location of the Song Position Pointer in the timeline of the sequencer. The Song Position Pointer, often called the *playhead* in other DAWs, indicates where Reason will start playback in the song. The top set of numbers indicates the song position in bars, beats, 1/16th notes, and ticks. The bottom set of numbers indicates the time position in hours, minutes, seconds, and milliseconds.

Like other DAWs, Reason divides the quarter note into 960 ticks. The Ticks field in the Song Position indicator shows the tick value relative to the 1/16th note increment, with 240 ticks available per 1/16th note.

Transport buttons. The transport buttons contain controls to rewind, fast-forward, stop, play, and record in the sequencer.

Loop Locator positions. The Loop Locator positions area indicates the position of the Left and Right Locators in the sequencer. These Locators define a range that Reason will play repeatedly when loop mode is active. The top number in the display indicates the Left Locator position (in bars, beats, 1/16th notes, and ticks); the bottom number indicates the Right Locator position.

 Click on the L or R button to the left of the Loop Locator position displays to move the Song Position Pointer to the Left or Right Locators, respectively.

System status indicators. The system status indicators area contains a small set of lights and meters that indicate aspects of the current status of Reason. Two of the most important are the CALC indicator, which lights up as Reason works on background audio calculations, and the DSP meter, which indicates Reason's CPU usage.

Main Mixer View

The Main Mixer view (shown in Figure 5.10) provides an environment for processing and mixing audio. This view displays a series of mixer channel strips. In general, you will have one channel strip that corresponds to each audio and instrument track in the Sequencer view, although Reason's powerful routing can enable other more complex scenarios.

Each strip includes controls for input gain and signal flow, dynamics processing (compression/expansion/gating), EQ, inserts, sends, panning, volume, and bus output assignment. The Main Mixer view also contains a master fader section that can process the final audio output of all of the channels after they are mixed together.

Figure 5.10 Main Mixer view in Reason Intro

To access the Main Mixer view when it is not displayed, choose **WINDOW > VIEW MAIN MIXER**.

 You can also display the Main Mixer view by pressing Function Key F5.

Input and Signal Path Controls. The top portion of each track's channel strip provides controls for adjusting the signal level coming into the mixer and setting the path the audio signal takes through the mixer channel.

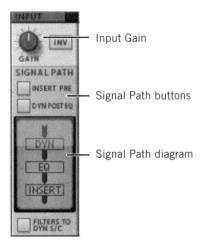

Figure 5.11 Input and Signal Path controls

Dynamics Controls. The second section of each track's channel strip provides controls for applying dynamics processing (compression and expansion/gating) to the audio flowing through the mixer. These processors can alter the dynamic range of an audio signal to help the volume of the audio sit well with the other tracks in a mix.

Figure 5.12 Dynamics controls

EQ Controls. The third section of each track's channel strip provides controls for applying EQ processing to the audio flowing through the mixer. The EQ adjusts the levels of certain frequencies within an audio signal. The EQ can be used to remove unnecessary or problematic frequencies in a sound or to boost frequencies you want to emphasize. This can also help the audio signal blend well with other tracks.

Figure 5.13 EQ controls

Insert Effects Controls. The fourth section of the mixer provides controls for managing insert effects. Inserts are audio processors that you add to process the signal as it flows through the mixer. Insert effects are generally configured through the Racks view. The Racks view also allows you to configure eight customizable "macro" controls (four rotary knobs and four buttons) for the effects, which are accessible from this section of the mixer channel strip.

Figure 5.14 Insert Effects controls

Send Effects Controls. The fifth section of the mixer provides controls for managing send effects. Sends make a copy of the audio signal flowing through the mixer channel and send it to audio effects that are connected to the mixer in a parallel path. Send effects allow you to send multiple audio signals from different channels to the same effect (such as a reverb unit). You can control the level of the audio signal being sent to the effect from each source channel. Reason's mixer supports eight send effects.

Figure 5.15 Send Effects controls

Fader Controls. The bottom portion of the mixer strip provides controls for setting playback options. These include Pan controls for positioning the signal within the stereo field; buttons to enable Mute and Solo functions; a Volume fader for setting the output level; and an Output selector for routing the audio to a mixer bus channel.

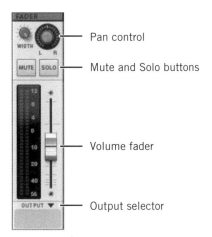

Pan control

Mute and Solo buttons

Volume fader

Output selector

Figure 5.16 Fader section controls

Racks View

The Racks view displays all of the devices that create and manipulate audio in your song. This view displays a series of vertical columns that contain the user interfaces for the devices. Reason displays devices in the Racks view as though they were virtual pieces of physical hardware in a studio rack. You can use the Racks view to create and delete devices and change the parameters of devices that control how they sound. VST plug-ins also appear as devices in the rack, although you need to open a window for a plug-in to see its controls and parameters.

A uniquely powerful feature of Reason is the ability to look at a back view of the rack to view and change the virtual cables that route audio signals to and from the devices in the rack.

Mix Channel Device

Player Device

Audio Track Device

Instrument Device

Browser

Effects Device

Figure 5.17 Racks view in Reason Intro

To access the Racks view when it is not displayed, choose **WINDOW > VIEW RACKS**.

 You can also display the Racks view by pressing Function Key F6.

Reason groups devices into a few different categories, as seen in the Browser, although all the different device types exist side by side in the rack.

Mix Channel Devices

A Mix Channel device is the Racks view representation of a channel strip in the Main Mixer view. The controls on the Mix Channel device in the Racks view, such as the Volume fader and the Mute/Solo buttons, mirror the controls for the Mix Channel in the Main Mixer view. Mix Channel devices receive audio from devices such as instruments and route the audio through the mixer so that it is represented in the Reason output. Most of the time, you will not need to create Mix Channel devices yourself, as Reason automatically creates them for you when you add instruments. However, you can also create additional Mix Channel devices manually. You can find the Mix Channel device under the Utilities section of the Browser.

Player Devices

Player devices can be thought of as MIDI effects. Rather than processing audio, these devices process note data coming from the sequencer. These devices manipulate the notes in some way before an instrument device plays the notes. You can find Player devices under the Players section of the Browser.

Audio Track Devices

An Audio Track device is the Racks view representation of an audio track in the sequencer. When you create an audio track in the sequencer, an Audio Track device is created in the rack. The opposite is also true: creating an Audio Track device in the rack creates an audio track in the sequencer. Audio Track devices also act like Mix Channel devices, and they have a corresponding channel strip in the Main Mixer view. As a result, Audio Track devices do not need to be routed to separate Mix Channel devices in order to be represented in the mix. Audio Track devices are found in the Utilities category of the Browser.

Instrument Devices

Instrument devices represent virtual instruments that receive MIDI note data from the sequencer (or from live input from a MIDI controller). These devices turn MIDI notes into audio that you can hear. Reason provides a wide variety of built-in instrument devices, found under the Instruments category in the Browser.

Effects Devices

Unlike Instrument devices, Effects devices don't create audio signals themselves. Instead, they receive audio from a source like an instrument device or an audio track and then process it in some way. Types of audio effects include reverb, chorus, EQ, distortion, and so on.

You can create effects devices by dragging them from the Browser. Simply drop the effect beneath the audio source device that you want to process, and Reason will connect the device into the signal flow. (For audio tracks, Reason will add the effect to the "insert effects" section of the track.) You can chain multiple effects together to create unique sounds!

Browser Panel in Racks View

The Browser looks and functions exactly the same in Racks view as in Sequencer view. You can use it to browse for devices available in Reason and drag devices into an empty area of the rack. You can also drag a device on top of another device of the same type to replace it. Unlike in Sequencer view, you cannot drag audio clips into the Rack view to create new audio tracks. However, you can browse your files and the built-in Reason sound libraries to find device *patches*—files that contain a stored representation of the parameters on a device that create a certain sound. You can drag a patch into the rack to automatically create the appropriate device and load the sound at the same time. For example, dragging a bass patch for the Thor synthesizer into the rack will create the synthesizer instrument and load the bass settings for you.

The Back of the Rack

An advanced and powerful feature of Reason is the ability to manipulate the signal flow among the devices in the rack by way of virtual patch cables connecting the devices. To see the back of the rack, simply press the **TAB** key. Another press of the **TAB** key will flip the rack back around to the front view.

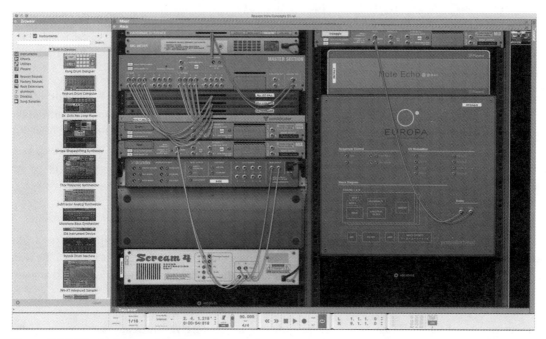

Figure 5.18 The back view of the Reason Intro rack

The back view of the devices in the rack reveals virtual jacks for routing audio input and output and control voltage (CV) signals between devices. Cables can be manipulated with simple drag-and-drop mouse actions by clicking on the jacks.

This view can be a bit intimidating when you first start working with Reason, but you generally won't need to use it right away; you can just create devices and tracks, and Reason will handle the cable routing for you behind the scenes. However, if you accidentally find yourself in this view or you want to try it out, it's important to have a general awareness of this interface.

Managing Views and Windows

Each of the main views uses a shortcut key to toggle the view on/off. These shortcuts are by far the most important keyboard commands in Reason, and you will likely use them many times while working on a session. Here is a summary recap of the shortcut keys.

Table 5.1 Reason main view key commands

To show this view...	...use this key:
Main Mixer view	F5
Racks view	F6
Sequencer view	F7

You can also use combinations of keys to show multiple views at the same time; for example, press **F5+F6** to show both the Main Mixer and the Racks views, or press all three function keys to show all the views at once.

By default, all of the views share the same window on your computer; using the keyboard shortcuts will switch the view that is currently visible. If you prefer to work with multiple windows (if you have a computer setup with multiple monitors, for example), you can also separate the views into their own separate windows using the **DETACH** arrow buttons in the interface. If you separate the views in this way, the keyboard commands will switch between the open windows.

Figure 5.19 The Detach buttons for the Main Mixer and Racks views

---

Function Keys on Mac Systems

On some Mac systems, you may find that the **F5, F6,** and **F7** function keys are mapped to operating system commands and do not work in Reason. Because switching Reason views is so important, it is recommended that you clear or reassign the operating system functions mapped to these keys so that you can use the keys with Reason. Please refer to the operating system documentation or the Reason documentation for details on configuring these keys.

Working with Tracks

Once you've created a project in Reason, you will need to create tracks for your project and give them meaningful names. Tracks are where your audio and MIDI performances are recorded and edited. Audio and MIDI data can be copied and duplicated in different locations to create repeating patterns, to arrange song sections, or to assemble material from multiple takes.

Adding Tracks

The method you use to add tracks in a Reason song varies depending on the type of track you want to create. Reason supports the following four main track types:

- **Audio**—This track type lets you record or import audio clips and work with them in a waveform-based display format.

- **Instrument**—This track type lets you record, create, and edit MIDI clips. Instrument tracks are linked to instrument devices in the Reason rack that play the MIDI note data.

- **Automation**—This track type is used with devices such as Effects devices that cannot play audio or MIDI clips. Automation tracks let you record, create, and edit automation clips that change parameters on the device in real time as the song plays. For example, you can use automation tracks to vary the amount of distortion being applied by a distortion effect device as your song plays back.

- **Transport**—This is a special track type that you cannot create or delete. Every song you create contains a single Transport track that is always located at the top of the sequencer. The Transport track holds automation clips that let you change the tempo and time signature of your song at different points.

Creating Audio Tracks

Reason provides many different methods to create audio tracks, including the following:

- From the main menu, choose **CREATE > CREATE AUDIO TRACK**, or press **COMMAND+T** on a Mac (**CTRL+T** on Windows).

- Drag an Audio Track device from the Utilities category of the Browser into the Racks view.

- Drag an Audio Track device from the Utilities category of the Browser into the Sequencer view.

- Click the **ADD TRACK** button in the Sequencer view below the Track List and choose the **CREATE AUDIO TRACK** option.

- Click the **ADD DEVICE** button in the Racks view below the existing devices and choose the **CREATE AUDIO TRACK** option.

- Drag an audio file from the Browser into an empty area of the Sequencer view or into an area that falls between two existing tracks.

Creating Instrument Tracks

Many DAWs require you to first create an empty instrument track and then assign an instrument to it, but Reason is a bit different. In Reason, you simply create an instrument device, and Reason automatically creates a corresponding track linked to the instrument. The easiest method to create instrument devices is to drag them from the Instruments category of the Browser into the Racks view or Sequencer view.

You can also initiate an instrument track using the **CREATE > CREATE INSTRUMENT** command or the shortcut **COMMAND+I** on a Mac (**Ctrl+I** on Windows). However, using this command simply displays the Browser focused on the Instruments category; you still end up creating the instrument from the Browser.

Creating Automation Tracks

To create an automation track, you first need to choose or create a device to automate. Automation tracks are used with Effects, Utilities, and Player devices that don't process audio or MIDI data. Unlike Instrument devices, when you create these device types, Reason does not automatically create sequencer tracks for them.

Once you choose a device to automate, you can create an automation track in a variety of ways. One technique is to right-click an empty area of the device interface in the Racks view and choose the **CREATE TRACK FOR <DEVICE NAME>** option near the bottom of the menu.

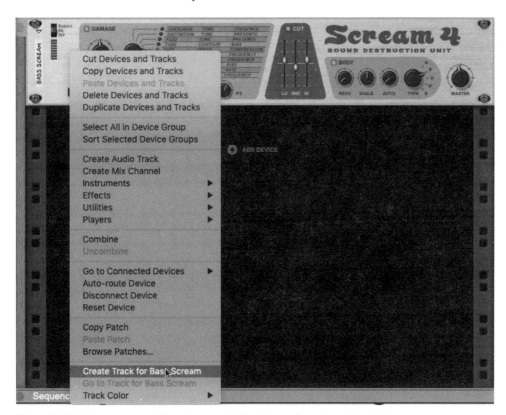

Figure 5.20 Creating a track for a Scream 4 effect device from the right-click menu

Figure 5.21 The empty automation track for the Scream 4 effect device in the Sequencer

Mono Versus Stereo Tracks

In Reason, you do not need to choose between mono or stereo format when creating audio and instrument tracks. Audio tracks automatically adapt according to how they are used. You can mix both mono and stereo audio clips on the same track. Instrument tracks in Reason similarly have no concept of mono or stereo format. The instrument devices in the rack can be connected to the main mixer through Mix Channel devices with mono or stereo output, but Reason will automatically make the correct connections for you when you create an instrument.

However, you will need to consider the distinction between mono and stereo when you are recording audio. You will need to choose whether an audio track will record a single mono input or a stereo input pair. For example, you will typically use mono input to record vocal and dialog performances, since the source signal will usually be a solo voice recorded with a single microphone. The same holds true for recording many types of close-miked acoustic instruments, as well as electric instruments that are connected directly to an audio interface. Common exceptions would include stereo-miked instruments, such as pianos, and electronic instruments that provide stereo outputs, such as drum machines and synthesizers.

Metronome Click

When using Reason to record music, it can be helpful to use a click. A click provides a metronome-based sound that performers can use as a reference tempo while recording. Reason has built-in click functionality and does not need a dedicated click track.

To turn the click sound on or off, simply toggle the **CLICK** button in the Transport Panel by clicking it with the mouse, or press the **C** key. You can use the vertical slider in the Transport Panel to adjust the volume of the click sound.

Figure 5.22 The click section of the Transport Panel

Edited Clips Versus Original Recordings

Each recording you complete on a track in Reason is stored in its entirety in your song file on disk. When you finish recording, the sequencer will contain a clip that includes all of the original, unedited audio.

When you edit a clip, such as by resizing the start or end of the clip or splitting the clip with the Razor Tool, the edited clips are pointers that reference a portion of the audio within the full recording.

By working with clips rather than entire recordings or individual audio files, Reason can perform nondestructive audio editing. This means the original recordings are unaltered by the edits you perform. As a result, you can recover or reference "missing" audio after an edit by using resizing and other techniques.

Use caution when deleting audio clips from your tracks. If you delete all audio clips that reference a specific recording, you will have no way to recover the original recording. In such a case, you can use the Undo function to recover the audio or revert to a backup copy of your song that contains the recording.

Tracks Versus Lanes

In the Reason Sequencer view, the specific part of a track that holds your clips in the Arrangement Pane is called a lane. In many cases, the distinction between a track and a lane is a minor one. Until you start working with automation, each of your tracks will usually have only one lane—a single row of audio clips for each audio track, and a single row of MIDI note clips for each instrument track.

However, the Reason Sequencer view supports multiple lanes of information for each track. Audio tracks will have only one lane of audio information, but you can have multiple lanes of automation clips associated with an audio track. Instrument tracks also support multiple lanes of automation clips, but they additionally support multiple lanes of note clips. This means that an instrument device could be playing notes from multiple clips at the same time!

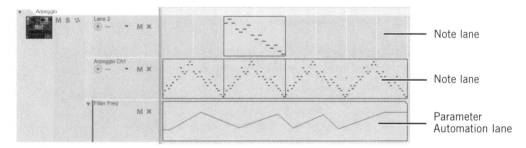

Figure 5.23 A single instrument track with multiple lanes

 Some Reason devices support yet another type of lane called a Pattern lane that can switch between different groups of settings stored on the device.

Again, to start out, you can think of tracks as having just one lane of clips. As you work with Reason, you will likely come across some operations that affect lanes as opposed to tracks, though, so it will be helpful to recognize the difference.

Basic Navigation

As a project grows, both in overall length and in track number, it becomes increasingly important to be able to navigate through the project quickly. In this section, we cover basic navigation techniques for use with playback, recording, and editing in a project.

Using the Transport Controls

The Transport Panel provides buttons for controlling playback and Song Position Pointer location.

Figure 5.24 Transport buttons in the Transport Panel

The transport buttons provide the following functions:

- **Rewind**—The Rewind button moves the Song Position Pointer backward one bar when clicked. You can click and hold the button to continue moving the song position back. This function is available both when the transport is stopped and while it is in motion.

- **Fast Forward**—The Fast Forward button moves the Song Position Pointer forward one bar when clicked. You can click and hold the button to continue moving the song position forward. This function is available both when the transport is stopped and while it is in motion.

- **Stop**—The Stop button stops a playback or record pass. The stop function can also be initiated by pressing the **SPACEBAR** when the transport is in motion.

- **Play**—The Play button initiates playback from the Song Position Pointer location. The play function can also be initiated by pressing the **SPACEBAR** when the transport is stopped.

- **Record**—The Record button begins a record pass and records content into any record-enabled track(s).

Zooming and Scrolling in the Sequencer View

As mentioned earlier in this chapter, you can use the Magnifying Glass Tool to zoom in on an area of the sequencer. Clicking on the sequencer with the Magnifying Glass Tool will zoom in horizontally and vertically for the entire Sequencer view. To reverse this behavior and zoom out, hold the **OPTION** modifier on Mac (or **CTRL** on Windows) while clicking with the Magnifying Glass Tool.

You can also zoom in and out horizontally by clicking and dragging a Song Navigator handle in the Song Navigator area in the Sequencer view. (See Figure 5.25.) The Song Navigator area looks like an enlarged horizontal scroll bar at the bottom of the Sequencer view. The shaded area between the rounded Song Navigator handles represents the time range displayed in the Arrangement Pane of the sequencer.

 Drag the right Song Navigator handle to quickly zoom in or out on displayed content. Drag the left handle to set the start of the displayed range. Click and drag on of the shaded area between handles to scroll the sequencer.

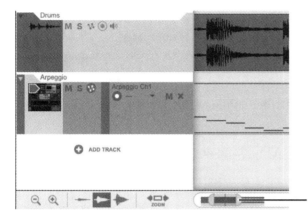

Song Navigator handle

Figure 5.25 Song Navigator handles in the Song Navigator area

Alternate Zoom Functions

Reason provides additional zoom functions that can speed up your workflow when you need to quickly zoom in and out. Available options include the following:

- **Zoom In**—To zoom in from the keyboard, press the **H** key. Each key press zooms in one level.

- **Zoom Out**—To zoom out from the keyboard, press the **G** key. Each key press zooms out one level.

- **Zoom to Selection**—To fit the currently selected clip or clips within the sequencer, zooming horizontally, press the **Z** key. Press **Z** again to zoom out horizontally to display all clips in the song.

 You can also activate Zoom to Selection by clicking the Zoom button to the left of the Song Navigator area at the bottom of the sequencer. (See Figure 5.25 above.)

- **Mouse Wheel Zoom**—When using a mouse with a scroll wheel, you can zoom in and out horizontally in the sequencer by holding **COMMAND+SHIFT** on Mac (or **CTRL+SHIFT** on Windows) while moving the wheel.

(i) With the Apple Magic Mouse, hold Command while swiping left/right to zoom horizontally. Holding Command while swiping up/down will zoom vertically.

Horizontal Scrolling

When zoomed in on your project, you will commonly need to scroll the Sequencer view left or right to access a desired location on the screen. The following options can be used to scroll the Sequencer view horizontally:

■ Drag left and right using the Song Navigator at the bottom of the Sequencer view. (Click within the shaded area in the Song Navigator.)

■ Click and drag left or right with the **HAND** tool to scroll horizontally.

■ Press the **. (PERIOD)** key on the numeric keypad to jump to the start of the song. You can also press **SHIFT+RETURN** twice to achieve the same effect.

■ When using a mouse with a scroll wheel, hold the **SHIFT** modifier and move the wheel to scroll horizontally.

You can also swipe left/right with a Magic Mouse, trackpad, or similar device to scroll the sequencer earlier or later.

Vertical Scrolling

When working on a project with many tracks (or with tracks set to large display sizes), you will commonly need to scroll the Sequencer view up or down to access a desired track on the screen. The following options can be used to scroll the Sequencer view vertically:

■ Drag up or down using the scroll bar on the right side of the view.

■ Click and drag up or down with the **HAND** tool to scroll vertically.

■ Press the **HOME** key on the numeric keypad to scroll to the top of the window.

■ Press the **END** key on the numeric keypad to scroll to the bottom of the window.

■ Press the **PAGE UP** key on the numeric keypad to scroll up by one screen.

■ Press the **PAGE DOWN** key on the numeric keypad to scroll down by one screen.

■ When using a mouse with a scroll wheel, turn the wheel with no modifiers to scroll vertically in the Sequencer view.

Controlling Playback Behavior

You can move the Song Position Pointer for playback using transport controls, such as fast-forward and rewind. However, you can also manually position the pointer anywhere within your project to set a playback location. You can also set the scrolling behavior that the Sequencer view uses during playback.

Setting a Playback Location

To set the playback location, do one of the following:

- Click anywhere on the Ruler. The Song Position Pointer will jump to the clicked location.

- Click and drag on the Ruler or directly on the Song Position Pointer to move the pointer to a new location.

(i) **If Snap mode is turned on, the pointer will stay locked to the grid and will move according to the chosen grid size.**

- Press the **[7]** key on the numeric keypad to move the pointer one bar earlier or the **[8]** key to move one bar later.

When you click the **PLAY** button in the Transport (or press the **SPACEBAR**), the song will begin playback from the current pointer location. The Song Position Pointer will move across the screen during playback, indicating the current playback point. Playback will continue until you press the **STOP** button (or press the **SPACEBAR** a second time).

The Follow Song Option

Reason provides a **FOLLOW SONG** option that lets you specify whether the contents of the Sequencer view will scroll automatically during playback and recording. When this option is disabled, the Song Position Pointer will move offscreen during playback, and the Sequencer view will not scroll.

When this option is enabled, the entire contents of the view will scroll by one screen each time the Song Position Pointer reaches the right edge of the screen, and the pointer will continue from the left edge of the view. In this mode, the screen will also jump to the Song Position Pointer whenever you start the transport or reposition the Song Position Pointer with key commands.

To toggle the scrolling behavior, choose **OPTIONS > FOLLOW SONG** or press the **F** key.

Monitoring Your Timeline Location

When navigating a project to set locations for playback, recording, or editing, you'll find it useful to reference the Song and Time Position indicators that Reason provides for you. You can refer to the indicators to determine the current Song Position Pointer location.

You change the current pointer location by clicking on one of the numbers in the Song/Time Position area of the Transport Panel and dragging up or down with your mouse. For example, you can click on the "beats" number in the Song Position display and drag the number to move the Song Position Pointer one beat at a time.

You can also double-click on one of the numbers in the Song/Time Position area of the Transport Panel to open a text box where you can type in an exact numeric location for the Song Position Pointer.

Making Selections

Making proper selections is an important part of any work you will do in a project. Selections in the Reason sequencer are entirely based on clips. Clips are selected by simply clicking them with the Selection Tool. A selected clip will have a distinct outlined appearance, and clip resize handles will be visible at the left and right edges of the clip. Selecting an audio clip will also reveal other controls for the clip, such as fade handles and a clip level handle.

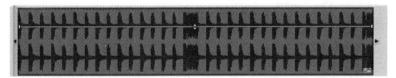

Figure 5.26 An audio clip selected in the sequencer

Unlike other DAWs, Reason does not let you select a portion of audio or MIDI data within a larger clip. If you want to isolate a portion of a clip for editing, you must first use the Razor Tool to make cuts in the larger clip and separate the area that you want to edit.

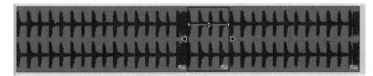

Figure 5.27 A smaller clip separated from a larger clip with the Razor tool

Review/Discussion Questions

1. Which view shows audio waveforms and MIDI performances in a timeline display on tracks and lanes in your Reason song? (See "Sequencer View" beginning on page 109.)

2. What are some of the sequencer tools and their functions? (See "Sequencer Tool Buttons" beginning on page 110.)

3. What do the Edit Mode buttons in the sequencer do? What are some of the available edit modes? (See "Edit Modes" beginning on page 111.)

4. What is the Ruler used for in Reason? What measurement unit is displayed on the Ruler in Reason? (See "The Ruler" beginning on page 113.)

5. What is the purpose of the Browser in Reason? How can it be used with the Sequencer view? (See "Browser Panel" beginning on page 114.)

6. What are some controls available in the Transport Panel? (See "Transport Panel" beginning on page 114.)

7. How are tracks displayed in the Main Mixer view? What are some controls available in this view? (See "Main Mixer View" beginning on page 115.)

8. Which view displays all of the devices that are used to create and manipulate audio in your song? (See "Racks View" beginning on page 120.)

9. How is the Browser used in the Racks view? (See "Browser Panel in Racks View" beginning on page 121.)

10. What is the purpose of the back view of the rack? What key can you press to flip between the front and back views of the rack? (See "The Back of the Rack" beginning on page 122.)

11. What track types are supported in Reason? What type of track lets you change the tempo or time signature of your song? (See "Adding Tracks" beginning on page 123.)

12. How is the process for creating an Instrument track different from that for creating an audio track? (See "Creating Instrument Tracks" beginning on page 124.)

13. What are clips in Reason? How is an audio clip different from a recording? (See "Edited Clips Versus Original Recordings" beginning on page 126.)

14. What are some ways to zoom in and out within the Sequencer view in Reason? What are some ways to scroll the Sequencer view to the left or right? (See "Zooming and Scrolling in the Sequencer View" beginning on page 128).

15. What are some methods to move the Song Position Pointer and set a playback location? (See "Setting a Playback Location" beginning on page 131.)

16. What position indicators does Reason provide to help you monitor your timeline location? (See "Monitoring Your Timeline Location" beginning on page 131.)

17. How can you make selections in the sequencer? Can you select just a portion of a clip? (See "Making Selections" beginning on page 132.)

 To review additional material from this chapter and prepare for certification, see the Reason Audio Production Basics Study Guide module available through the Elements|ED online learning platform at ElementsED.com.

Configuring and Working on a Song

🎧 Activity

In this exercise, you will configure the display for the project you created in Exercise 2. You will also practice making selections to inspect clips. By using zoom functions and navigating in the Sequencer view, you will be able to accurately focus on specific sections of the song. You will then configure grid settings and create a looped playback section over a portion of the song.

🕐 Duration

This exercise should take approximately 15 minutes to complete.

⊕ Goals/Targets

- Adjust the working area of the Sequencer view
- Select clips with the Selection Tool
- Zoom and navigate the Sequencer view
- Use Snap mode and set the grid size
- Configure loop locators and loop a section of a song

Exercise Media

This exercise uses media files taken from the song, "Overboard," provided courtesy of The Pinder Brothers, along with remix files provided by Eric Kuehnl.

The media provided for this course may be used for educational purposes only. No rights are granted to use the media for any other personal, commercial, or non-commercial purposes.

* The mix, processing, and media files have been adapted for use in the exercises contained herein.

Getting Started

To get started, you will open the multitrack project you created in Exercise 2. This will serve as the starting point for this exercise.

 This exercise can be completed using Reason Intro, the standard full version of Reason, or Reason Suite.

Open your existing Overboard project:

1. Launch your Reason software, if it isn't already running. Once startup completes, a new song window will display.

2. Open the **FILE** menu. Your project may be visible in the list of recent songs at the bottom of the menu; if so, just click the file name to open it. Otherwise, continue with the remaining steps.

3. Click the **OPEN** command from the **FILE** menu. The Browser will be displayed in the current song window. It will switch to the Open Song context.

4. Use the Browser to navigate to the location where your Overboard01-xxx song is stored.

 Refer to Exercise 2 for information on navigating files and folders in the Browser.

5. Select the Overboard01-xxx song and click the **OPEN** button at the bottom of the Browser to open the file. Or, simply double-click on the file in Browser.

6. The song will open as it was when last saved.

Configuring the Sequencer View

Before starting work on a project, it can be helpful to adjust the view in Reason to optimize the window displays. In this section of the exercise, you will adjust the display of the Sequencer view, set an initial zoom level, and hide other panels for an unobstructed layout.

Configure the onscreen display for the Sequencer view:

1. If the Sequencer view is not already visible in full screen, choose **WINDOW > VIEW SEQUENCER** or press **F7** to show only the Sequencer view.

2. To get an overview of the entire song, press the **Z** key or click the **ZOOM** button at the bottom of the Sequencer view. Because all of the clips in the song extend for the full length, it won't matter if you

have a clip selected when you use the command. The Sequencer will zoom horizontally to fit all of the content on screen.

3. If the Browser is visible, choose **WINDOW > HIDE BROWSER** or press **F3** to toggle the display.

4. If the Tutorial Panel is visible, choose **WINDOW > HIDE TUTORIAL** to hide the panel by collapsing it on the right side of the screen.

Making Selections and Loops

The tracks in this session represent a segment of a remix of the song "Overboard" by the Pinder Brothers. To get familiar with the audio in this project, you will start by selecting a clip and examining its duration.

Select audio on the Kick track:

1. Activate the **SELECTION TOOL** in the Sequencer view toolbar if it is not already active.

Figure 5.28 Selection Tool active in the Sequencer view toolbar

2. Click on the clip in the **01 Kick** track's audio lane to select it.

Figure 5.29 Selection on the 01 Kick track

3. Refer to the Inspector area in the Sequencer view toolbar to determine the length of your clip in bars and beats. In the example shown below, the clip starts at Bar 1, Beat 1 (1.1.1.0), and is just over 39 bars long (39.0.0.64).

Figure 5.30 Clip details in the Inspector area of the Sequencer view toolbar

Adjust the zoom for a better view:

1. Use one of the following methods to zoom in a few levels:

    *   Activate the **MAGNIFYING GLASS TOOL** and click on the track lane.

Figure 5.31 Magnifying Glass Tool positioned in the middle of the 01 Kick track lane

- Press the **H** key on the keyboard.

2. Zoom in sufficiently to see each bar number on the Ruler.

> (i) **Try zooming by dragging a Song Navigator handle or by holding Command+Shift (Mac) or Ctrl+Shift (Windows) while scrolling with a mouse scroll wheel.**

3. Use the horizontal scrolling method of your choice to reach the end of the selected clip on the 01 Kick track in the Sequencer view.

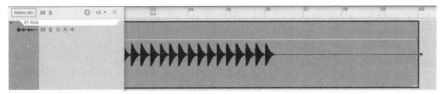

Figure 5.32 Close-up view of the end of the clip on the 01 Kick track

4. Press the **Z** key or click the **ZOOM** button in the Transport Panel to zoom back out and see the full length of the clip in the Sequencer view.

Set up loop playback for the guitar solo:

1. Listen through the song to determine where the guitar solo occurs.

2. You should hear the guitar solo begin around Bar 19 and continue through around Bar 27.

3. Identify the Left Locator in the Ruler, as follows:

- Look at the Loop Locator positions area in the Transport Panel. The values displayed here specify the song position of the Left and Right Locators.

```
L    1. 1. 1.   0  ⇕
R    9. 1. 1.   0  ⇕
```

Figure 5.33 Loop Locator positions area in the Transport Panel (Left Locator at Bar 1, Right Locator at Bar 9)

- Using the position information for the Left Locator, scroll the Sequencer view as necessary until you see the Left Locator in the Ruler area above the tracks; it will be a pointer marked with an **L**. (See Figure 5.34.)

(i) **Clicking on the L or R button in the Loop Locator positions area will scroll to the Left or Right Locator, respectively, if the Follow Song option is enabled.**

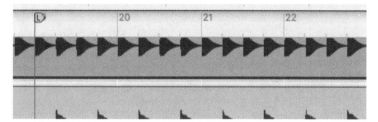

Figure 5.34 The Left Locator in the Ruler

4. Enable and configure the grid, as follows:

- In the Sequencer view toolbar, enable Snap mode, if needed, by clicking the **SNAP** button or by pressing the **S** key. The Snap button will be shaded when active.

- Set the grid size to 1 bar by clicking the drop-down menu and choosing **BAR**.

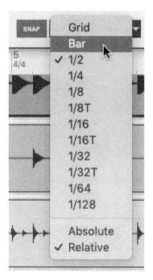

Figure 5.35 Selecting a one-bar grid size

5. Click on the Left Locator in the Ruler and drag it to Bar 19. With Snap mode active and a one-bar grid, the locator will stay locked to the grid as you move it and will move in whole-bar increments.

Alternatively, you can click and drag the loop placement numbers up and down to change the location of the Loop Locator, click one of the numbers and use the up/down arrows to change the values, or double-click a number, type an exact position, and then press **ENTER** to move the locator.

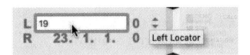

Figure 5.36 Typing in a location for the Left Locator

6. Set the Right Locator to Bar 27 using the method of your choice. The Right Locator is a pointer marked with an **R** in the Ruler.

7. Enable Loop mode by clicking the **LOOP ON/OFF** button in the Transport Panel so that it becomes enabled (shaded).

   You can also toggle Loop mode on/off by pressing the L key.

Figure 5.37 The Loop On/Off button in the Transport Panel

8. Click on the **L** button in the Loop Locator area to move the Song Position Pointer to the Left Locator.

When you are done, your Sequencer view will look similar to Figure 5.38.

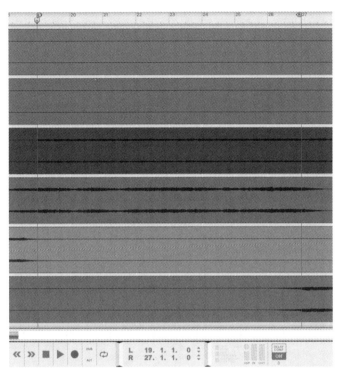

Figure 5.38 Loop configured over Bars 19 to 27 of the guitar solo

9. Press **SPACEBAR** to listen to the loop range. When the Song Position Pointer reaches Bar 27, it will loop back to Bar 19 and continue playing (repeating the guitar solo continuously).

Finishing Up

Feel free to experiment further with the project as time allows. Practice making selections and navigating in the Sequencer view. Also try setting the grid to different values to make loops that start and end in between bars. When finished, return to the start of the project and save the work you've done.

Save the project changes:

1. Click the **STOP** button in the Transport controls twice or press the **[.] (PERIOD)** key on the numeric keypad to place the Song Position Pointer at the start of the song.

2. Choose **FILE > SAVE** to save the song.

3. Choose **FILE > CLOSE** to close the song.

That completes this exercise.

Chapter 6

Reason Concepts, Part 2

...What You Need to Know to Work with Reason Software...

We begin this chapter with an overview of recording audio and MIDI performances using Reason Intro. Next, we explore options for importing existing audio and MIDI data into your song. And finally, we take a look at basic editing techniques that can be used to fine-tune audio and MIDI performances.

Learning Targets for This Chapter

- Learn how to record audio

- Learn basic MIDI recording techniques

- Learn how to import audio and MIDI

- Understand how the Snap function affects edit operations

- Learn basic techniques for working with clips

Key topics from this chapter are illustrated in the Reason Audio Production Basics Study Guide module available through the Elements|ED online learning platform. Sign up at ElementsED.com.

When starting a new Reason Intro song, the first step is typically to either record or import some audio or MIDI data into your session. This can be accomplished in a number of ways, but generally you'll either record new clips directly onto an audio or instrument track in your project or import existing clips from various locations on your computer.

Setting Up for Recording

When beginning a music project, it can be beneficial to set the song tempo for Reason prior to starting to record. This will allow you to use the click as a tempo reference during recording, helping the musicians keep the performance consistently in time. Additionally, recording to a click will ensure that the music aligns to the Ruler. As a result, any edits you make can easily be aligned to the music by working on the grid with the Snap function activated.

 The song tempo defaults to 120 beats per minute (BPM) in Reason software.

To set the song tempo, do the following:

1. Find the tempo value in the Transport Panel at the bottom of the Reason Intro window.

Figure 6.1 The song tempo in the Transport Panel

2. Do one of the following:

    •   Click on the number and drag it up or down.

    •   Click the **TAP** button repeatedly at the tempo you'd like to use. Reason Intro will average the tempo as you click.

    •   Double-click the tempo number, which will open a text box where you can type in a new value and then press **RETURN** or **ENTER** to set the tempo.

To use a click with the song, do one of the following:

■   Click on the **CLICK** button in the Transport Panel with the mouse.

■   Press the **C** key to toggle the click on or off.

Recording Audio

Recording audio is one of the most fundamental aspects of working with any modern DAW. Working in Reason Intro gives you nonlinear access to all of the audio in your project. This means you can jump directly to any location without needing to fast-forward or rewind like you would with a tape recorder.

Reason Intro provides a truly intuitive workflow for quickly getting your musical ideas recorded onto a track. And thanks to the large storage drives included with most modern computers, you'll rarely have to worry about running out of recording space.

Setting a Record Location

Before you begin recording, you'll need to specify where you want the record pass to begin. Audio engineers call the beginning of the target area the *punch-in point*. The punch-in point in Reason Intro is designated in the same way that you set the playback location: by moving the Song Position Pointer to the desired spot on the Ruler. Reason Intro does not provide the concept of a punch-out point or record region, where a record pass automatically stops upon reaching a designated end point.

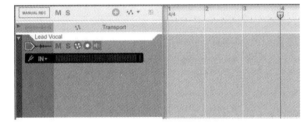

Figure 6.2 The Song Position Pointer set to record at Bar 4 on a lead vocal track

To move the Song Position Pointer, use any of the techniques described in previous chapters, such as clicking or dragging in the Ruler, or changing the placement numbers in the Transport Panel.

Record-Enabling Audio Tracks

Once you've set a location for recording, you'll need to specify which tracks to record on. To specify a destination track, you need to record-enable the target track in the Sequencer view.

With the default settings in Reason Intro, you can record-enable a track in the Sequencer view by clicking on the track to select it. You can click anywhere on the blank space of a track header, as seen in Figure 6.3, to select the track. Doing so will automatically move the Master Keyboard Input indicator to the track, which will in turn record-enable the track.

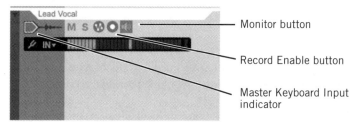

Monitor button

Record Enable button

Master Keyboard Input
indicator

Figure 6.3 A record-enabled track in the Sequencer view

Only one track in the sequencer can have the Master Keyboard Input focus at a time, so as you click on different tracks to select them, the previously selected track will no longer be record-enabled; in other words, only one track will be record-enabled at a time by selecting. To record-enable additional tracks, click the Record Enable buttons on each of the desired tracks.

If a track is record-enabled, you can click the Record Enable button to take the track out of record-enable mode. Turning off record-enable mode in this way works regardless of whether the track was record-enabled by being selected or manually with the button.

Monitoring Audio Tracks

An important function that goes along with record-enabling tracks is *monitoring* tracks. This means listening to the live input that is coming into the track from the source you are recording, such as your microphone, guitar, or other instrument.

The default settings of Reason Intro make track monitoring easy. Reason Intro will automatically activate monitoring whenever you record-enable a track, and because Reason Intro record-enables a track when you select it, monitoring will also be enabled at the same time just by selecting the track.

Monitoring allows you to hear the track input while the transport is stopped and also while you are recording. When you click **PLAY** (or press the **SPACEBAR**), Reason Intro will disable monitoring and you will instead hear any clips already in place on your tracks. Tracks that have no clips will remain silent. This is true even when a track is record-enabled and you begin playback: you will hear any existing clips on that track and *not* the live input from your audio interface.

You can manually control monitoring with the **MONITOR** button in the Sequencer view. (Refer to Figure 6.3.) You can click the button to turn monitoring on or off to override the default behavior of Reason Intro. The monitoring setting for a track is independent of the record-enable setting, so you could, for example, monitor a track that is not record-enabled, or start playback and turn on monitoring for a track to still listen to the live input while playing your song. Once you manually toggle monitoring, though, Reason Intro will no longer automatically control it.

This can be a little confusing, so ultimately if you're not hearing what you expect from a track, check the monitoring button to verify the track settings.

Avoiding Feedback

Reason software's default settings make working with audio tracks easy. Selecting a track record-enables the track and begins monitoring it, so you can hear the incoming audio signal and start recording quickly. Unfortunately, this functionality also creates an opportunity to accidentally create audio feedback noise.

If you have a microphone connected to your audio interface when you create or select an audio track using the microphone input, the audio track immediately begins monitoring the microphone input. With Reason outputting to speakers, any microphone that is close to the speakers or sensitive enough to pick up their output creates a risk of feedback.

This problem is particularly noticeable if you are using the built-in speakers and microphone on a laptop. Because the speakers and microphone are naturally close to one another, you will probably encounter harsh feedback noise almost immediately.

When possible, make sure to position your microphone away from your speakers, set the microphone input gain to a safe level, reduce the track output level in the Mixer view, or listen on headphones so that the microphone can't pick up and feed back the audio output. In the case of a laptop where you have limited control over the microphone and speakers, you may want to change Reason's Audio Preferences settings to prevent automatic monitoring of record-enabled tracks.

Selecting Audio Track Input

Before recording audio to a track, you will need to choose the input or input pair on your audio interface that is routed to the track. For example, if you are recording a microphone connected to an input on your audio interface, you will need to select that input as the source for the audio track.

Reason software provides two locations where you can select the input for an audio track. In the Sequencer view, after you record-enable an audio track, the track header in the Track List will display a selector labeled IN that provides the audio input options for the track. (See Figure 6.4.) In the Racks view, the corresponding Audio Track device contains a selector labeled **AUDIO INPUT** that provides the audio input options for the track. (See Figure 6.5.) You don't need to record-enable the track to see the selector in the Racks view, but you may need to click the **FOLD/UNFOLD** button (triangle) for the device to expand it to view the input selector.

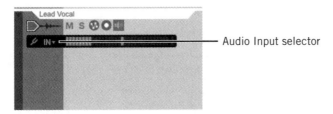

Figure 6.4 Audio Input selector in the Sequencer view

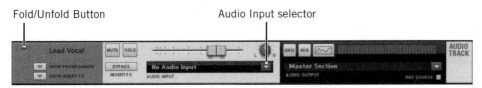

Figure 6.5 Audio Input drop-down list in the Racks view

Both methods of accessing the audio input list display the same set of options. The top part of the list contains options for Mono Input and Stereo Input. Choosing **MONO INPUT** will configure the audio track to record a single, mono audio signal from one input of your audio interface. Choosing **STEREO INPUT** will configure the audio track to record a pair of audio signals from two inputs of your audio interface, making a stereo audio signal.

Below the mono and stereo options, Reason lists the available inputs or input pairs available on your audio interface. You can set the audio input or stereo input pair to use for a track by selecting it in the list.

Initiating a Record Take

Once you have record-enabled one or more tracks, you can initiate a record take in Reason. The most common technique is to press the **RECORD** button (red circle) in the Transport Panel.

Figure 6.6 The transport controls in the Transport Panel

Reason Intro also provides a number of keyboard shortcuts that can be used to initiate recording. If you record frequently, these are some of the first shortcuts you'll want to learn!

To begin recording, do one of the following:

- Press **COMMAND+RETURN** (Mac) or **CTRL+RETURN** (Windows).

- Press the **[*]** key on the numeric keypad.

Stopping a Record Take

The record take will continue until you manually stop the transport. Reason Intro provides several techniques for stopping a record take.

To immediately stop a record take, do one of the following:

■   Press the **STOP** button in the transport controls.

■   Press the **SPACEBAR**.

■   Press the **[0]** key on the numeric keypad.

■   Press **SHIFT+RETURN**.

■   Press the **[*]** key on the numeric keypad. In this case, recording will disengage, but the transport will continue moving in playback.

Settings That Affect Recording and Monitoring

Reason software includes a few settings in Reason that you may want to change, either to fit your preferred workflow or to avoid accidentally recording or monitoring tracks that you didn't intend to affect.

In the upper left of the Track List you'll find a button labeled **MANUAL REC**. Activating this button prevents Reason from automatically record-enabling tracks as you select them. With this button active, you must use the Record Enable buttons for individual tracks to change their record state, giving you more control and preventing accidental recording on tracks. You can toggle **MANUAL REC** on/off at any time to change this behavior as needed.

Manual Rec button

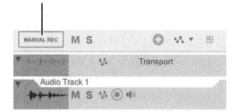

Figure 6.7 The Manual Rec button at the top of the Sequencer view Track List

In Reason's preferences screen under the **AUDIO** tab you'll find a section labeled Monitoring. This section has three options to control audio track monitoring. The default setting is Automatic, which allows Reason to turn on monitoring for record-enabled tracks, as described above. The second option, Manual, is useful to prevent automatic monitoring; in this mode Reason requires you to use the track Monitor buttons to choose tracks you want to monitor. This gives you more control and may help to avoid unexpected microphone feedback. The last setting, External, disables all monitoring and is intended for situations where you are monitoring signals externally using a dedicated mixer or other hardware device.

Recording MIDI

Although the MIDI communication protocol is more than 30 years old, understanding how to record and edit MIDI data is more relevant today than ever. Advances in virtual instrument design and computer processing speed over the past decade have made virtual instruments an indispensable part of the modern music production workflow. The fundamental workflow for recording MIDI is similar to that for recording audio. However, some important differences apply in MIDI production. But don't worry; you don't need a PhD in synthesis to start making music with the MIDI feature set in Reason Intro!

Instrument Creation and Routing

The monitoring workflow is one area that differs significantly when recording MIDI versus recording audio. This is because an audio signal can be monitored directly, whereas MIDI data must be routed to an instrument to become an audible signal. The routing can seem a little complicated at first, but once you understand the basics, you'll be able to take advantage of this powerful technology.

 For basic information on routing MIDI data to virtual instruments, see "Tracking with Virtual Instruments" in Chapter 4 and "Creating Instrument Tracks" in Chapter 5.

To recap some of the discussion from Chapter 5, you can start working with virtual instruments in Reason by using the Browser to access the Instruments category. From there, you can drag an instrument into the Racks or Sequencer view to add the instrument device to your song. You can also double-click an instrument to add it. Alternatively, instead of creating an instrument device directly, you can use the Browser to find device "patches," which you can then drag into the Sequencer view or Racks view to create an instrument with its settings preconfigured for a particular sound.

Whether you create an instrument directly or by loading a patch, Reason automatically sets up the necessary routing for you. Understanding the signal flow of MIDI and audio data with the instrument device can be helpful when working in Reason. Creating an instrument causes Reason to add a track for the instrument in the Sequencer view. The track sends note data from clips in the track's lanes to the instrument. Instruments and instrument tracks in Reason have a one-to-one relationship: an instrument can have only a single instrument track in the sequencer, and an instrument track in the sequencer is linked to only that one instrument device in the Racks view.

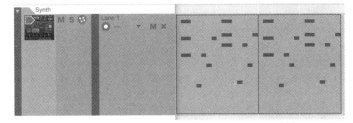

Figure 6.8 An instrument track in the sequencer with a lane containing note clips

The Racks view allows you to access the parameters on a device that control the way it sounds. The instrument device receives MIDI note information and creates an audio signal that you can hear.

Figure 6.9 An instrument device in the Racks view

For you to hear the audio signal coming from the instrument device, the signal needs to pass through the main mixer and then ultimately out through your audio interface and output device. To connect the audio output of the instrument device to the main mixer, Reason uses a utility device called a Mix Channel. The Mix Channel creates a channel strip in the main mixer that can receive audio from a device such as an instrument. Reason software will automatically create and connect a Mix Channel device when you create an instrument device, but understanding the signal flow is important.

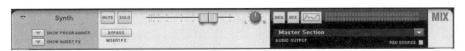

Figure 6.10 A Mix Channel device in the Racks view

An instrument device connects to a Mix Channel device with virtual cables leading from the audio output of the instrument to the audio input of the Mix Channel. Again, when you create an instrument, Reason automatically makes this connection for you, so you normally don't have to do anything with it. However, if you want to see or change the connections between your devices, you can do so from the back view of the rack. As a reminder, you can switch between the back and front views of the rack by pressing the **TAB** key.

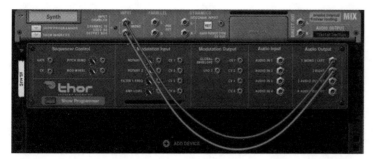

Figure 6.11 Cables connecting an instrument to a Mix Channel in the Racks view

The Mix Channel device receives the audio signal from the instrument. The Mix Channel can process the audio signal and route it through the mixer, combining it with the other tracks in your song. The combined audio signal then passes through your audio interface to a listening device, such as your speakers or headphones. You can see and control the Mix Channel device as a channel strip in the Main Mixer view.

Figure 6.12 A portion of the mixer channel strip for a Mix Channel device shown in the Main Mixer view

When you put all of these pieces together, you get a complete signal flow from the MIDI notes on the instrument track, to an audio signal generated by an instrument device, to a Mix Channel device and the main mixer, and then ultimately to sound that you can hear.

Playing Instruments

Once you create an instrument, if you have a keyboard or other MIDI controller attached to your computer, you'll probably want to start playing the instrument. The Master Keyboard Input function applies to instrument tracks as well as audio tracks. This control determines which instrument track has focus and receives the notes that you play on a connected controller. Using Reason's default settings, you can move the Master Keyboard Input to an instrument track by selecting it (clicking on it) in the Sequencer view, just as with audio tracks.

Master Keyboard Input
indicator

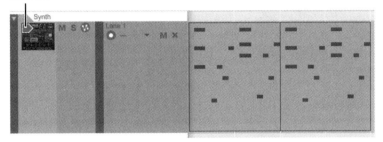

Figure 6.13 An instrument track with Master Keyboard Input active

In many cases, you will be working in the Racks view while configuring and adjusting the sounds of your instruments. In those cases, it can be inconvenient to switch to the Sequencer view to change the Master Keyboard Input target. Fortunately, the Racks view also allows you to change Master Keyboard Input so that you can play different instruments.

To enable the Master Keyboard Input for a device, click on the instrument device to select it in the Racks view. The device will be surrounded by a selection border, and the upper-left corner of the border will display a Master Keyboard Input tab control. Click on this control to assign Master Keyboard Input to the selected instrument. This has the same effect as moving the selection in the Sequencer view and will allow you to play the instrument.

Master Keyboard
Input tab

Figure 6.14 A selected instrument in the Racks view

Record-Enabling Instrument Tracks

Record-enabling instrument tracks is quite similar to recording-enabling audio tracks. In fact, as you change the Master Keyboard Input to play different instruments, you will also be record-enabling the instrument tracks. Just as with audio tracks, the default settings for Reason automatically record-enable an instrument track as you select it in the Sequencer view. Similarly, changing the Master Keyboard Input in the Racks view selects the corresponding track, which in turn record-enables it.

Like audio tracks, instrument tracks have Record Enable buttons in the Sequencer view that you can click to manually turn record-enable mode on or off. However, there are a few important differences between audio tracks and instrument tracks in this area.

First, unlike audio tracks, which have only a single lane for audio data, instrument tracks can have multiple lanes of note data. Rather than having a single Record Enable button for an entire instrument track, each lane has a Record Enable button so that you can specify the lane you want to record on. Only one lane on an instrument track can be record-enabled at a time. In most cases, your instrument tracks will have only one lane, making this distinction less important.

Second, while it's possible to easily set up multiple audio tracks for recording by clicking their Record Enable buttons, the same is not true for instrument tracks. Only a single instrument track can be record-enabled at a time. Clicking the Record Enable button for an instrument track's note lane will turn off record-enable mode for any other instrument track lane.

 It is possible to play and record multiple instrument tracks at the same time in Reason, but doing so requires a more advanced feature called Surface Locking.

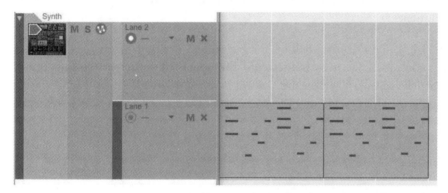

Figure 6.15 An instrument track with two note lanes; Lane 2 (top) is record-enabled, as indicated by the red circle with a white center.

Finally, instrument tracks do not have Monitor buttons as audio tracks do. When an instrument track has Master Keyboard Input, it will always output the notes that you play on your MIDI controller, regardless of whether the transport is stopped, playing, or recording. No additional mode needs to be activated to hear the audio generated by instrument devices.

Merging and Layering MIDI Performances

In Reason, when you record on top of a note clip that you've previously recorded, successive passes add MIDI data, merging it into the existing clip. This process is great for recording the left-hand and right-hand parts of a piano performance separately, for example, and combining them both in the same clip.

Merging MIDI performances is also a helpful tool for recording drum parts. You can start by recording just one or two drum pieces in the first pass (such as kick and snare) and then merge in additional pieces on subsequent passes. (See Figures 6.16 and 6.17 below.)

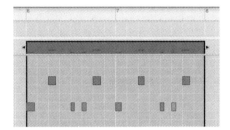

Figure 6.16 The first pass of a drum recording with just kick and snare

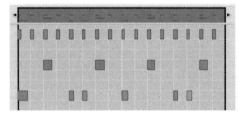

Figure 6.17 The second pass of a drum recording with hi-hat merged into the clip

Instead of merging MIDI data in this way, you can take advantage of Reason's support for multiple note lanes in a single instrument track to record the different drum pieces to separate clips on separate lanes. To do so, you would start by recording the first set of drum pieces, such as kick and snare, as above. Then, create a new note lane for the instrument track.

To create a new note lane, do one of the following:

■   Select the instrument track and click the **NEW NOTE LANE** button at the top of the Track List in the Sequencer view.

New Note Lane button

Figure 6.18 The New Note Lane button

■   Select the instrument track and click the **DUB** button in the Transport Panel.

Figure 6.19 The Dub and Alt buttons in the Transport Panel

 The Alt button is similar to the Dub button, but it mutes the original note lane on the track as it adds a new lane.

You can use this process to record another drum piece, such as the hi-hat, onto a separate clip in the new note lane. The new clip layers on top of the original kick and snare clip and plays simultaneously with it so that you hear all of the drum parts together. Your track will look similar to Figure 6.20.

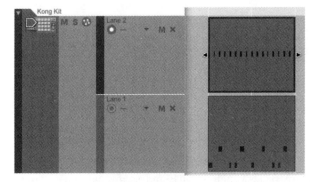

Figure 6.20 A layered drum track with two note lanes

Importing Audio and MIDI

If you're not ready to record in Reason Intro, or if you'd like to use existing media from your hard drive instead, you can import audio or MIDI files into your Reason song to get started. There are a number of ways to import audio and MIDI files into Reason.

Supported Audio Files

Before you attempt to import an audio file, you'll want to make sure that it is in a format that can be imported into Reason Intro.

All of the following audio file types can be imported into Reason Intro:

- AAC
- AIFF

- MP3

- ReCycle (REX files)

- WAV

- M4A

- WMA (Windows only)

- CAF (Mac only)

 Refer to the Operation Manual for your Reason product for complete details on supported file types.

Importing from the Desktop

An easy way to get started importing is to use the drag-and-drop method. Reason Intro fully supports drag-and-drop importing of audio and MIDI files from the Mac Finder or Windows Explorer. Simply select the desired files on your computer and drag them into your Reason project. (We look at the different destinations available for dropped files below.)

Importing from the Browser

Another way to import audio and MIDI files into Reason is to use the built-in Browser. In addition to providing access to Reason's devices, the Browser provides functions similar to a Mac Finder or Windows Explorer window. You can use the Browser to navigate and search any mounted volume on your computer. However, the Browser offers features that are tailored to working with audio and MIDI files.

To open the Browser if is not already visible, do one of the following:

- Select **WINDOW > SHOW BROWSER**.

- Press Function Key **F3.**

The Browser will open. (See Figure 6.21.)

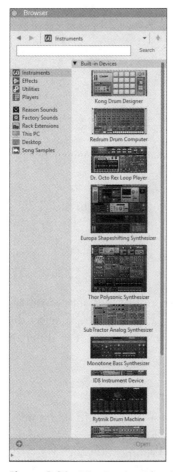

Figure 6.21 The Browser showing the Locations List on the left and the Browser List on the right (Windows system pictured)

Typically, the Browser is used in one of two ways: for manually browsing or searching for files.

To manually browse for files in the Browser:

■ Use the shortcuts in the Locations List on the left to browse any of the following:

- Built-in sound libraries (Reason Sounds, Factory Sounds)

- Libraries that are bundled with Rack Extension plug-ins

- The contents of your computer (This PC and Desktop on Windows, your user home directory and desktop on macOS)

- Audio clips recorded or stored as samples using Reason's sampling features (Song Samples)

- Custom Locations or Favorites Lists that have been added to the Browser

Figure 6.22 The Locations List in the Browser

To search for content using standard browser functionality:

1. Select an item in the Locations List on the left side of the Browser where you want to start the search.

2. Click on the text box at the top of the browser and type in a search term.

3. Press **ENTER** or click the **SEARCH** button to begin a search.

About Browse Focus

At times, the Browser will enter a mode called "browse focus" where it narrows down the types of files that it will show and links the Browser to a particular device or track. Browse focus is indicated by an orange background in the Browse Focus field at the top of the Browser.

Browse patches for: Cosmic Waves ⊗

◀ ▶ | 📁 Pads ▼ | ⬆

Figure 6.23 An active browse focus in the Browser

A common example of this occurs when you create a new instrument device. Reason sets browse focus for the device so that you see instrument patches that you can access to load a particular sound. The browse focus context links the Browser to the instrument, so that you can double-click a patch in the Browser to load it in the instrument device.

However, if you want to then use the Browser to find other types of files, you may not be able to see them while the Browser has an active browse focus. In that case, click the **X** button in the Browse Focus field at the top of the Browser to clear the current browse focus.

File Drag and Drop Locations

Once you've located a file on your desktop or using the Browser, you'll want to consider your options for adding the file into a Reason song. Imported audio files can be dragged to the Sequencer view and dropped on the Track List, a blank area of the Arrangement Pane, or an existing audio track lane in the Arrangement Pane. Imported MIDI files can only be dropped onto the Arrangement Pane of the Sequencer view.

To import an audio file to the Track List:

1. Select an audio file in the Browser, the Mac Finder, or Windows Explorer.

2. Drag and drop the file into the Track List. A new track containing the imported audio data will be added to the project.

To import an audio file to a new track in the Arrangement Pane:

1. Select an audio file in the Browser, the Mac Finder, or Windows Explorer.

2. Do one of the following:

    *   Drag and drop the file onto the empty space at the bottom of the Arrangement Pane.

    *   Drag and drop the file between two existing tracks in the Arrangement Pane. Reason will display a line with a + symbol indicating that the audio will be added to a new track at location of your mouse cursor.

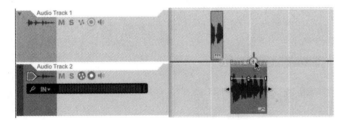

Figure 6.24 Dropping an audio file between existing tracks to create a new track

A new track containing the imported audio data will be added to the project.

To import an audio file to an existing track:

1. Select an audio file in the Browser, the Mac Finder, or Windows Explorer.

2. Drag and drop the file to the desired location on the audio lane of an existing track in the Arrangement Pane.

To import a MIDI file:

1. Select a MIDI file in the Browser, the Mac Finder, or Windows Explorer.

2. Drag and drop the file to the desired location in the Arrangement Pane. New tracks will be created containing the imported MIDI data.

You cannot drop MIDI files onto existing tracks.

Using the Import Command

Audio and MIDI files can also be imported into Reason Intro using the FILE > IMPORT AUDIO FILE and FILE > IMPORT MIDI FILE commands, respectively. However, these commands do not use a basic file dialog box. Instead, they simply show the Browser with a corresponding browse focus for importing audio or MIDI files. The actual import process must then be completed by dragging and dropping from the Browser or by selecting files and using the IMPORT button at the bottom of the Browser.

Selecting files and clicking the IMPORT button will place the files at the current Song Position Pointer location.

The Snap Function and the Grid

Before you start editing clips, it helps to have a good understanding of the Snap function, as well as the grid values that are available in Reason Intro. The Snap function will impact the movement and placement of clips and the usage of various sequencer tools. You should also know how to modify the grid value so that clips will align properly when working with the Snap function activated.

The Snap and Grid Controls

As discussed in Chapter 5, the Snap function and the grid settings are found in the Sequencer view toolbar. You can toggle the Snap function by clicking the SNAP button or by pressing the S key. You can select a grid value by clicking on the selector next to the Snap button.

Figure 6.25 The grid controls in the Sequencer view toolbar (Snap function active)

Working Without the Snap Function

When the Snap function is not active, Reason places no constraints on how clips can be moved and edited. Clips can be moved to any location on a track, even on top of another clip. (The right-most clip with the latest start time will be the audible one.)

Working with the Snap Function

Enabling the Snap function offers you much of the same freedom to move and edit clips on tracks. However, with the Snap function active, clips will align to grid intervals when moved, as specified by the grid value. Edits that you make by resizing clips or using tools such as the Razor Tool will lock to the grid automatically. The Snap function makes editing a song very easy, because all of your edits can occur on a bar or beat.

Setting the Grid Value

When working with the Snap function, you'll want to set the grid to an appropriate size. Use the grid value selector in the Sequencer view toolbar to set the grid size as desired.

Figure 6.26 The grid value selector in the Sequencer view toolbar

The available grid values range from **1 bar** to **1/128 note**, with triplet options denoted by a T, for 1/8 note triplets, 1/16 note triplets, and 1/32 note triplets.

The grid value selector contains an additional item at the top simply called **Grid**. Choosing this option activates an adaptive grid mode. In this mode, the grid value is based on your horizontal zoom setting in the Sequencer view. As you zoom out, the grid value automatically gets larger, moving toward one bar in size. As you zoom in, the grid value automatically gets smaller, moving toward 1/128 note in size. See Figure 6.27 below for an illustration of adaptive grid mode.

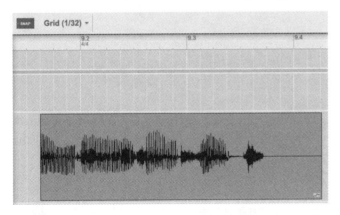 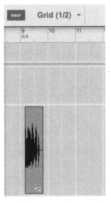

Figure 6.27 While zoomed in on an audio clip in adaptive grid mode, the grid value adjusts to a 1/32 note size (left). After zooming out, the grid value adjusts to a half-note size (right).

Working with Clips

Once you've recorded or imported some audio or MIDI material, it's time to start editing!

Selecting a Sequencer Tool

To get started editing, you'll need to select a sequencer tool in Reason Intro. You can select tools in several ways. One obvious method is to click directly on the desired tool in the toolbar. However, it can often be faster to use keyboard shortcuts.

Figure 6.28 The available sequencer tools in the Reason Intro Sequencer view

The keyboard shortcuts for the sequencer tools are pretty easy to remember. They simply use keys **Q** through **I** (the top row of letters on the keyboard) from left to right. These are among the first shortcuts you should learn in Reason.

- To enable the Selection Tool, press **Q** on your computer keyboard.

- To enable the Pencil Tool, press **W**.

- To enable the Eraser Tool, press **E**.

- To enable the Razor Tool, press **R**.

- To enable the Mute Tool, press **T**.

- To enable the Magnifying Glass Tool, press **Y**.

- To enable the Hand Tool, press **U**.

- To enable the Speaker Tool (when available), press **I**.

Basic Editing Techniques

DAWs provide a variety of ways that you can slice and dice audio and MIDI clips. You'll become familiar with a multitude of editing techniques in Reason Intro as you work, but for now we'll cover the basic techniques you'll need to get started.

The basic editing techniques include the following:

- Selecting clips

- Moving clips

■   Cutting, copying, and pasting clips

Let's go through each of these basic building blocks in some detail.

Making Selections

Selecting material is one of the most basic and frequently used techniques in the DAW world. It serves as a foundation for further editing. Many commands require a clip to be selected first before they can be executed. In Reason Intro, you can make selections in a number of ways.

To select a single clip:

■   Click on the clip with the **SELECTION** Tool.

To select multiple clips, do one of the following:

■   Using the **SELECTION** Tool, click an empty area of the Arrangement Pane and drag to draw a rectangular selection region. Release the mouse button to select all of the clips within the region.

■   Using the **SELECTION** Tool, hold the **CTRL** modifier (Windows) or the **SHIFT** modifier (Mac) and click on clips one at a time. Each clip you click will be added to the set of currently selected clips.

 Selections do not directly include blank areas between clips. However, edit operations like cut and paste will preserve the relative spacing of clips and blank areas between them as appropriate when multiple clips are selected.

Moving Clips

At times you may want to move a clip to a new location on a track or onto a completely different track. While you can use Cut/Copy/Paste for this purpose, it's usually faster to simply drag the clip to the desired location using the **SELECTION** Tool.

To move a clip to a new location on the same track:

■   Using the **SELECTION** Tool, click and drag the clip horizontally.

To move a clip to a different track:

■   Using the **SELECTION** Tool, click and drag the clip vertically.

 When dragging clips to a different track, you can hold Shift to constrain the clips and prevent them from moving left or right. Press the Shift key after clicking the mouse to begin the movement process; otherwise, if you hold Shift first, you may inadvertently change which clips are selected instead.

When moving clips to different tracks, you should move the clips between track lanes of the same type. For example, move audio clips onto another audio lane, and move note clips to another note lane. If you try to move a clip onto a different type of lane, you probably will not get a useful result. In most cases, the clip will become what is known in Reason as an "alien clip." For example, if you drag a MIDI note clip onto an audio lane, the note clip becomes alienated and will not be able to play back on the audio lane.

Figure 6.29 An alien note clip denoted by red vertical lines on an audio lane to the right of an audio clip

Cutting, Copying, and Pasting Clips

For edits involving many clips, you'll often want to use the Cut/Copy/Paste commands. These commands can be accessed at the top of the Edit menu, but you'll definitely want to use the keyboard shortcuts to save time. Reason Intro uses standard keyboard shortcuts for each of these operations:

- Cut operation—**COMMAND+X** (Mac) or **CTRL+X** (Windows)

- Copy operation—**COMMAND+C** (Mac) or **CTRL+C** (Windows)

- Paste operation—**COMMAND+V** (Mac) or **CTRL+V** (Windows)

Advanced Editing Techniques

Beyond selecting and moving your audio or MIDI data, you'll often need to use slightly more advanced editing techniques. Some useful options include resizing the start/end of a clip, splitting a clip into two or more separate clips, and adding fades to clips. All of these operations use nondestructive editing in Reason.

Nondestructive Audio Editing

Nondestructive editing is an import concept to understand when editing audio data in a DAW. Essentially, nondestructive editing means that any changes you make to an audio clip in Reason Intro will not affect the original audio file stored on disk or the original recording stored with your song. This gives you the freedom to try any number of different edits without fear of permanently modifying the original audio.

 Although you can edit audio nondestructively, if you delete all of the clips that reference a particular recording, you will not be able to recover the recording without using the Undo command or reverting to an earlier copy of your song file.

Editing MIDI Clips

While audio editing is nondestructive, the same is not true when editing MIDI data. This is because MIDI clips do not use the concept of pointing to an underlying recording as audio clips do. As a result, most

MIDI editing functions *will* directly modify a clip. Although it is possible to use the undo command to revert a clip to its original state, this option is typically useful only immediately after an unwanted change.

Resizing the Start and End of Clips

Resizing operations are used constantly when editing audio in Reason Intro. Resizing will essentially add or remove audio from the beginning (head) or end (tail) of a clip.

Resizing allows you to make a clip shorter by removing audio or MIDI material from the start or end of the clip. You can also make a shortened audio clip longer again at *any time* in the future by resizing it back out. In other words, resized clips still have access to *all* of the original audio data from the recording. This lets you recover any missing audio at a later time.

> (i) If you resize an audio clip beyond the boundary of the original recording, the extended parts of the clip will be silent.

> (i) In Reason, shortening a clip to remove part of the original audio recording is known as "masking" the recording.

To mask the start of a clip:

1. Use any technique to select the clip.

2. Using the **SELECTION** Tool, click and hold the Clip Resize handle near the clip start.

3. While continuing to hold the mouse, drag to the right to remove audio and move the clip start later.

Figure 6.30 Dialog clip before resizing (left) and after resizing (right)

To mask the end of a clip:

1. Use any technique to select the clip.

2. Using the **SELECTION** Tool, click and hold the Clip Resize handle near the clip end.

3. While continuing to hold the mouse, drag to the left to remove audio and move the clip end earlier.

If you have multiple clips selected, they will all be resized simultaneously when you resize any one of the clips.

Splitting a Clip into Two or More Clips

Another very common editing technique is to split a clip into two or more smaller clips. In Reason Intro, you can use the Razor Tool to perform this function.

 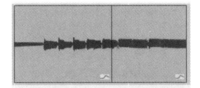

Figure 6.31 A single clip before splitting (left) and the result of splitting with the Razor Tool (right)

To split a clip into two clips:

1. Activate the **RAZOR TOOL** in the Sequencer view toolbar.

2. Click at the desired location inside a clip to split the clip at that position.

 The Razor Tool works with the Snap function. If the Snap function is turned on when you click to split a clip, the clip will be split at the grid position nearest to where you click.

Fading Clips

Last but certainly not least, you can apply fades to your audio clips. Fades are commonly used to gradually fade audio in and out; however, they can also be used to prevent undesirable pops and clicks at clip boundaries. Here, we'll focus on the basics of creating and editing fades in Reason Intro.

To create a fade in or fade out:

1. Select a clip to reveal the Fade In and Fade Out handles for the clip.

Fade In handle   Fade Out handle

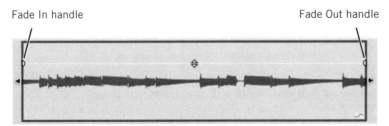

Figure 6.32 A selected clip with Fade In and Fade Out handles

2. Click and drag the Fade In handle to the right to add a fade-in or click and drag the Fade Out handle to the left to add a fade-out.

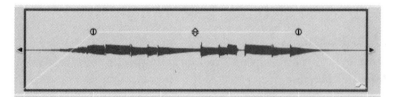

Figure 6.33 A selected clip with a fade-in and fade-out applied

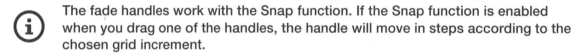

The fade handles work with the Snap function. If the Snap function is enabled when you drag one of the handles, the handle will move in steps according to the chosen grid increment.

In addition to fade-ins and fade-outs, you can create crossfades between adjacent clips. Crossfades are useful to smooth out an edit, by fading out the audio from before the edit point while simultaneously fading in the audio from after the edit point.

To create a crossfade:

1. Use the **SELECTION TOOL** to move two clips so that they partially overlap.

2. Use the **SELECTION TOOL** to select the right-most of the two clips (the clip with the later starting position).

3. Right-click the selected clip and choose **CROSSFADE** from the bottom of the pop-up menu or press the **X** key to enable the crossfade.

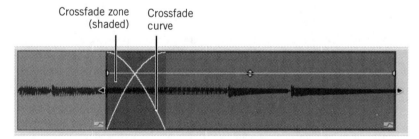

Figure 6.34 A crossfade applied to two overlapping clips

If you decide to modify the fades later, you can use a few different techniques depending on the type of fade you want to modify. To modify a fade-in or fade-out, select a clip and drag the corresponding fade handle to a new position.

To modify a crossfade, do one of the following:

■ Click on an edge of the crossfade zone with the **SELECTION TOOL** and drag it left or right to resize the crossfade. This will resize one of the overlapping clips.

■ Move one of the overlapping clips to correspondingly move and resize the crossfade.

■ Click on the crossfade curve at its center point in the crossfade zone and drag it left or right to change the shape of the crossfade.

To remove a crossfade, do the following:

1. Use the **SELECTION TOOL** to select the right-most of the overlapping clips on which the crossfade was originally enabled.

2. Right-click the selected clip and choose **CROSSFADE** again in the pop-up menu to deactivate it or press the **X** key to disable the crossfade.

 Moving the crossfaded clips apart so that they no longer overlap will also cause the crossfade to disappear. However, the crossfade will not be deactivated and will reappear if the clips are positioned to overlap again.

Review/Discussion Questions

1. How can you specify the starting point (punch-in point) for recording? (See "Setting a Record Location" beginning on page 145.)

2. How do you record-enable an audio track? (See "Record-Enabling Audio Tracks" beginning on page 145.)

3. What are two ways that you can begin a record take in Reason Intro? (See "Initiating a Record Take" beginning on page 148.)

4. What are two ways that you can immediately stop a record take? (See "Stopping a Record Take" beginning on page 149.)

5. How does the audio output of an instrument device connect to the main mixer? (See "Instrument Creation and Routing" beginning on page 150.)

6. How can you select an instrument to play with your MIDI controller? (See "Playing Instruments" beginning on page 152.)

7. How is merging or layering MIDI performances useful when recording drum parts? (See "Merging and Layering MIDI Performances" beginning on page 154.)

8. What are some types of audio files that can be imported into Reason Intro? (See "Supported Audio Files" beginning on page 156.)

9. What three locations are available as destinations for audio files you import using the drag-and-drop method? What location is available as a destination for MIDI files you import using the drag-and-drop method? (See "File Drag and Drop Locations" beginning on page 159.)

10. What does the Snap function do? (See "The Snap Function and the Grid" beginning on page 161.)

11. Which keys on the computer keyboard can be used to select the sequencer tools? (See "Selecting a Sequencer Tool" beginning on page 163.)

12. Which sequencer tool must be used to drag and drop a clip to a new location in the song? (See "Moving Clips" beginning on page 164.)

13. What happens if you move a clip onto a different type of lane that does not support the clip? (See "Moving Clips" beginning on page 164.)

14. Is audio editing destructive or nondestructive? What about MIDI editing? (See "Advanced Editing Techniques" beginning on page 165.)

15. What sequencer tool is used to split a clip into two separate clips? (See "Splitting a Clip into Two or More Clips" beginning on page 167.)

 To review additional material from this chapter and prepare for certification, see the Reason Audio Production Basics Study Guide module available through the Elements|ED online learning platform at ElementsED.com.

Importing and Editing Clips

🎧 Activity

In this exercise, you will learn how to perform basic editing techniques in Reason software. You'll begin by creating a new song. Then you'll import audio files and a MIDI file into the project and configure the tracks. Finally, you'll do some basic audio editing to remove unnecessary content from audio clips.

🕒 Duration

This exercise should take approximately 15 minutes to complete.

✦ Goals/Targets

- Create a new blank song document

- Import audio and MIDI into the project

- Assign the NN-XT sampler instrument for the MIDI parts

- Remove audio at the beginning of two drum tracks for a four-bar intro

---

Exercise Media

This exercise uses media files taken from the song, "Lights," provided courtesy of Bay Area band Fotograf.

Written by: Zack Vieira and Eric Kuehnl; Performed by: Fotograf

The media provided for this course may be used for educational purposes only. No rights are granted to use the media for any other personal, commercial, or non-commercial purposes.

Getting Started

To get started, you will create a new song to use for the exercise.

Create the song:

1.   Launch your Reason software to create a new song. If Reason is already running, create a new song by selecting FILE > NEW or pressing COMMAND+N (Mac) or CTRL+N (Windows).

2.   Choose WINDOW > VIEW SEQUENCER, if needed, to bring the Sequencer view to the forefront and make it active.

3.   Maximize or resize the window as needed for a full-screen view.

Importing Audio

In Exercise 2, you used the Browser to import audio files into your first project. Here, you'll use a similar technique to start a second project. You'll be using this new project for the remaining exercises in this book.

Import the audio files into your project:

1.   Press SHIFT+RETURN (Mac) or SHIFT+ENTER (Windows) two times to ensure that the Song Position Pointer is at the beginning of the song.

2.   Choose FILE > IMPORT AUDIO FILE to display the Browser and set the browse context to Import Audio File.

3.   In the Browser, navigate to your Documents folder (or other location where you saved the Reason APB Media Files folder in Exercise 1).

4.   Open the 06. Lights folder within the Reason APB Media Files folder.

5.   Click the Sort Arrow at the top of the file list, as needed, to display the files in descending numerical order.

6.   Select all ten WAV audio files in the folder. Reason may start auto-auditioning the files in the Browser, depending on your settings.

7.   If needed, click the Stop button (square icon) or the Auto button at the bottom of the Browser to cancel the audition.

8.   Click IMPORT to complete the import. Ten tracks will be added to your project, starting from the bottom of the list, and the imported audio will be placed on the tracks.

When finished, your project should look similar to Figure 6.35 below.

Figure 6.35 An overview of the imported audio

Importing MIDI

In this part of the exercise, you will import MIDI clips to two instrument tracks using a Standard MIDI file.

Import the MIDI data:

1. Choose FILE > IMPORT MIDI FILE. The Browser will display set to Import MIDI File.

2. In the Browser, locate and select the Exercise-06.mid file inside the 06. Lights folder.

3. Click IMPORT to import the file, or double-click the file to import it. Several MIDI clips will automatically be placed on two instrument tracks at the correct locations in the project.

Rename the tracks:

▪ Rename each of the instrument tracks to simplify their names as follows:

   • Double-click on the name of the Piano Ch1 track to open a text box for changing the name.

Figure 6.36 Renaming the Piano Ch1 track

   • Modify the track name to remove the "Ch1" ending and press ENTER or RETURN.

   • Repeat the process for the Strings Ch1 track.

Assigning Virtual Instruments

With the MIDI clips imported to instrument tracks, you'll next need to replace the virtual instruments on those tracks. When the Reason software imported the MIDI data, it created two ID8 instrument devices and set them to default piano sounds. For this song, you will change the instrument tracks to use different sounds that work well alongside the audio tracks. You'll use the NN-XT sampler virtual instrument on both tracks and load appropriate patches.

Replace the ID8 devices on the instrument tracks:

1. Choose **WINDOW > VIEW RACKS** to activate the Racks view and bring it to the forefront.

2. Locate the two ID8 instrument devices named Piano and Strings in the rack.

3. Choose **WINDOW > SHOW BROWSER** to display the Browser, if it isn't already visible.

4. Click on the **FACTORY SOUNDS** library in the Locations List in the Browser.

5. Navigate through the folders in the library by double-clicking them to open them. For the Piano track, open these folders in order: **NN-XT SAMPLER PATCHES > PIANO**.

6. In the list of patches displayed in the Browser List, scroll down to find the file called MK II Dynamic.sxt.

7. Click on the patch file and drag it from the Browser. While still holding the mouse button, move your mouse over the ID8 instrument for the Piano track. Reason will highlight the ID8 device with an orange overlay indicating that the device will be replaced.

8. Release the mouse button to place the patch file. The ID8 instrument device will be replaced with an NN-XT instrument device. The device will be configured with the MK II Dynamic.sxt sound loaded.

Figure 6.37 Dragging a patch from the Browser to replace an ID8 instrument device

9. Repeat the above process starting from Step 4 for the **Strings** track. This time, open **FACTORY SOUNDS > NN-XT SAMPLER PATCHES > STRINGS**.

10. Drag the **Stereotron.sxt** patch onto the Strings ID8 instrument device to replace it with an NN-XT instrument device with the **Stereotron.sxt** sound loaded. Your Piano and Strings instrument devices in the rack should now look similar to the Figure 6.38.

Figure 6.38 Two NN-XT instruments configured for the Piano and Strings tracks

Preview the project:

1. Choose **WINDOW > VIEW SEQUENCER** or press **F7** to toggle back to the Sequencer view.

2. Turn off the Loop function, if needed, by clicking the **LOOP ON/OFF** button in the Transport Panel so that it becomes disabled (unshaded).

3. Press the **SPACEBAR** to begin playback and listen to the project in its current state.

4. Pay particular attention to the number and type of instruments that you hear. Feel free to solo and mute tracks to isolate certain elements and get familiar with the different parts.

Removing Unwanted Audio

In this part of the exercise, you will remove some drum parts at the beginning of the song. This will create a more dramatic intro featuring the Synth Bass part.

Remove the unwanted audio:

1. Press the **H** key to zoom in two or three levels for a better view of the 01 Kick and 02 Snare tracks.

2. Enable the Snap function, if needed, by clicking the **SNAP** button in the Sequencer view toolbar or pressing the **S** key.

3. Set the grid value to one bar, if needed, using the grid value selector next to the Snap button.

Figure 6.39 Setting the grid value to one bar

4. Using the **SELECTION TOOL**, click on the clip in the **01 Kick** track audio lane to select it.

5. Using the **SELECTION TOOL**, click on the Clip Resize handle for the selected clip and drag it to the right, removing the first four bars of audio, so that the clip starts at Bar 5. You can verify the new position of the clip using the Ruler, or the Inspector in the Sequencer view toolbar.

6. Repeat Steps 4 and 5 for the **02 Snare** track.

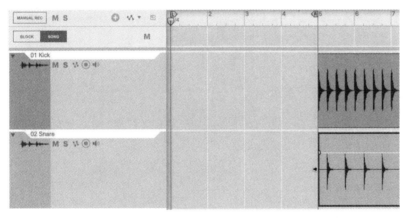

Figure 6.40 The first four bars removed from the Snare and Kick tracks

Finishing Up

To complete this exercise, you will need to save your work and close the project. You will be reusing this project in Exercise 7, so it's important to save the work you've done.

Before you wrap up, you should also listen to the project to hear the changes you've made.

Finish your work:

1. Choose **FILE > SAVE** to save the song. Name the song **Lights** and add your initials to the end of the filename.

2.   Press **[.]** on the numeric keypad to move the Song Position Pointer to the beginning of the project.

3.   Press the **SPACEBAR** to begin playback. Only the synth bass should be audible for the first four bars, creating a nice intro. Listen through the intro and the next several bars.

4.   When finished, press the **SPACEBAR** a second time to stop playback.

5.   Choose **FILE > CLOSE** to close the song.

That completes this exercise.

Chapter 7

Mixing Concepts

...What You Need to Know to Mix a Project...

This chapter introduces you to mixing concepts, from setting levels and panning to the role of EQ and dynamics processing on the tracks in a mix. We discuss how to keep a mix from clipping, the difference between gain-based processing and time-based processing, and the differences between insert processing and send-and-return processing. The chapter ends with a discussion of mixing in the box and some suggestions for maximizing your results when mixing in Reason.

✦ Learning Targets for This Chapter

- Set effective levels for your tracks

- Recognize the role of EQ and dynamics processing in creating a balanced mix

- Use panning to create a sense of space and positioning in your mix

- Understand how inserts and sends can be used to add processing to tracks

- Recognize the advantages of in-the-box mixing

 Key topics from this chapter are illustrated in the Reason Audio Production Basics Study Guide module available through the Elements|ED online learning platform. Sign up at ElementsED.com.

After you have recorded, imported, or otherwise created the media that you want to use in your project, you can set about the task of mixing the audio. Mixing is the process of setting the basic levels, positioning, and sonic characteristics of your tracks. Essentially, this is where you determine how the various parts of the project will blend together to create a cohesive result during playback.

Basic Mixing

Although mixes can get very complex on large sessions, the basic process is fairly simple. The primary goal when creating a stereo mix is to set the levels for each of the tracks using the tracks' Volume faders and to position each of the tracks within the stereo field using the tracks' pan controls. Other options include using inserts and sends to process each track and add effects or using the built-in EQ and dynamics effects in the Reason mixer.

Setting Levels

Setting the levels for your tracks is typically done in the Main Mixer view. Here you'll find Volume faders for each track, which you can use to adjust each track's overall output level.

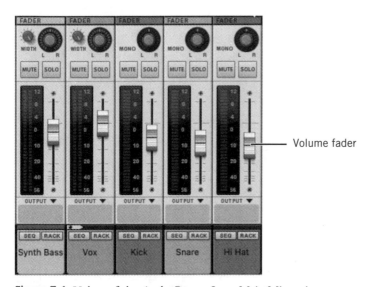

Volume fader

Figure 7.1 Volume faders in the Reason Intro Main Mixer view

Level Considerations

The purpose of adjusting the output level of a track is to make the track audible without obscuring other tracks or causing the track to become overly prominent in the mix. During the recording process, audio is often recorded louder than it needs to be in the final mix. Additionally, as you begin summing multiple tracks together, the overall output level for the project will increase. For these reasons, it helps to pull all of the faders down quite a bit when you begin mixing and then gradually adjust the levels of each track as needed.

Allowing one track to be heard above all the others is not only a matter of raising its Volume fader, however. A common mixing dilemma is that making one track louder will cause another equally important track to become overshadowed and indistinct.

Getting a Good Music Mix Is More Than Setting Levels

Raising the fader on a guitar track enough to get it to cut through the mix can make the vocal track hard to hear. Raising the fader on the vocal track may then begin to obscure the impact of the drums. And raising the faders on the drum tracks can affect the bass guitar, which may now be barely audible. By the time you bring up the level on the bass guitar to compensate, you may be right back where you started, with the guitar track no longer cutting through the mix. Only now, everything is louder!

In the quest for the ultimate mix, often you simply end up with a very loud mix and still cannot distinguish individual parts.

As you start setting the initial levels for your mix, try to achieve a good overall balance. Don't get overly concerned if you find that certain tracks start to compete with one another. These issues can be addressed later using panning, EQ and dynamics processing, and other techniques that help give each track its own unique space in the mix.

Beware of Clipping

Among the problems caused by loud tracks, the output levels of your mix may get too "hot," leading the signal to clip at the digital-to-analog converters. This will cause the mix to become distorted on playback during the loudest moments.

Digital distortion is always detrimental to sound quality and should be avoided at all costs. Therefore, it is better to mix too quiet than it is to mix too loud. You can always increase the overall output levels of a mixed stereo file at a later stage if needed. But you cannot remove clipping from the stereo file after the fact.

Metering on Source Tracks

To help you keep an eye on your audio levels, each track includes a standard meter display to the left of the track fader. Meters for individual track channel strips in Reason Intro are designed to imitate VU meters found on traditional analog equipment. These types of meters are different from digital peak meters. They react a bit more slowly to changes in the audio signal and give you a view of the average amplitude of the signal. The measurements on these meters do *not* represent amplitude in decibels relative to full-scale audio (dBFS), where 0 designates the maximum possible level; instead, 0 is found closer to the middle of the meter in Reason. Levels reaching above 0 in a track VU meter will display in red to indicate a high-level signal.

A common misconception in Reason is that a red track VU meter indicates a track that is clipping, but this is not the case. Because of the high bit depth and enormous dynamic range used internally by the Reason mixer, it's nearly impossible to clip an individual track during playback. However, you still need to prevent clipping when recording audio and when playing back the full mix!

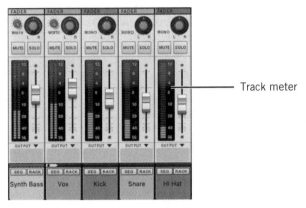

Track meter

Figure 7.2 Meters on tracks in Reason Intro

Unlike some DAWs, the track meters in Reason show *post-fader* metering. This means that the meter reflects the output level of the track rather than the level from the source audio or instrument. A red meter on a track indicates that the output levels for the track are too hot and can cause distortion if not attenuated in the Master Section. In this case, you can lower the track's Volume fader to reduce the playback amplitude. However, not having a red meter does not guarantee that the track will play back cleanly: if the combination off all of your tracks is too loud, the overall mix output levels could be pushed into clipping. This won't be shown on the individual track meters.

Track metering reflects effects processing done on the track, including changes made by EQ or dynamics processors. Any boost or cut to the signal caused by effects will be represented on the meter.

Metering in the Master Section

To ensure that the output levels of your mix are not too hot, it's a good idea to learn how to use the metering in the Master Section of the Main Mixer view. The Master Section can be used to monitor and control the output levels for your song. The Master Section and its associated Volume Fader always appear on the right side of the Main Mixer view.

The Volume Fader in the Master Section is sometimes called the Master Fader. Other DAWs may refer to the entire Master channel strip as the Master Fader.

The level meter next to the Volume Fader in the Master Section reflects the overall output of the entire mix. This is a summed total of the signals from each contributing track's mixer channel.

Clip indicators at the top of the meter in the Master Section will light red when the summed signal level exceeds the capabilities of the digital-to-analog converters. In this case, you can reduce the levels on each of the contributing channels in your mix or reduce the overall output level using the Volume Fader in the Master Section itself. Once any clipping has occurred, the clip indicators will remain red until you clear them by clicking the **RESET** button.

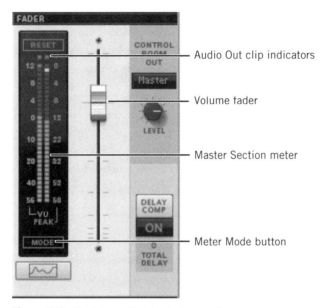

— Audio Out clip indicators

— Volume fader

— Master Section meter

— Meter Mode button

Figure 7.3 The Master Section in Reason Intro

While working on your mix, keep an eye on the meter in the Master Section. If the meter begins clipping, reduce the output levels before continuing. Unlike meters for individual tracks in the Main Mixer, the Master Section meter supports several meter modes that provide different ways of measuring the level of your mix. You can change modes by clicking the **MODE** button at the bottom of the meter.

 Hovering over the Mode button with your mouse will show a tooltip that indicates the current mode.

Reason Intro provides the following meter modes for the Master Section:

- **VU**: In VU mode, the Master Meter behaves identically to the individual track meters, showing an average view of the mix level, and not indicating peak levels.

- **VU + Peak**: In VU + Peak mode, the Master Meter acts primarily as a VU meter. However, this mode adds a peak level segment display in the form of an additional set of lights in the meter above the VU level that indicate the digital peak level of the audio. In this mode, the blue numbers to the left of the meter indicate the values for the main VU level, while the orange numbers to the right of the meter indicate the values in dBFS for the peak level. Note that the orange peak numbers place 0 dBFS at the top of the meter, indicating the point where digital clipping can occur.

■ **PPM**: In PPM mode, the Master Meter rises nearly instantaneously in response to increases in the signal level but takes a bit of time to drop back down when the signal level decreases. The result is that you can get a sense of the general level of the transient peaks in the mix. The PPM mode measures the signal in dBFS with zero (clipping) at the top. PPM stands for Peak Program Meter.

■ **PPM + Peak**: In PPM + Peak mode, the Master Meter acts primarily as a PPM meter. However, this mode adds a separate peak level segment display in the form of an additional set of lights in the meter above the PPM level that provide another indication of the dBFS peak level of the audio.

■ **Peak**: In Peak mode, the Master Meter both rises and falls nearly instantaneously in response to changes in the signal level, providing the most accurate metering display available. Peak mode is often the most useful meter mode, but the rapid movement of the meter may make it hard to read. If you run into that issue, try one of the PPM modes instead. The Peak meter measures the audio signal in dBFS, with zero (clipping) at the top. This mode also has a separate peak level segment display that lights up above the main meter bars. The peak segment display has some hold time before it drops down, giving you a view of the highest peak reached in a recent window of time.

The Role of EQ and Dynamics Processing

As mentioned above, setting the fader levels is not the only consideration when it comes to allowing each track to be heard in a mix. Other important considerations include the track's frequency spectrum and the track's dynamic response.

The frequency spectrum of a track, or the amount of energy the track has at different audio frequencies, can be shaped using equalization (or EQ for short). You can use EQ processing to help reduce low-end rumble, high-end hiss, and other unwanted noise in a signal. You can also use EQ to shape a signal and create a unique tone for each track. This allows certain target frequencies or frequency ranges to be more prominent than others. By selectively boosting and cutting frequencies on each track, you can balance the audio spectrum in the mix and give each track its own sonic footprint.

 In the interest of keeping track levels in check, it is generally better to focus on cutting the frequencies you don't want rather than boosting the frequencies you wish to emphasize.

The dynamic response of a track refers to the range of amplitude values on the track. Said another way, the track's dynamics relate to the differences in loudness from one part of the track to another. The momentary loudness peaks (or *transients*) on a track are often much louder than the track's average levels. These loudness peaks can begin to obscure other tracks (or cause clipping) long before the source track's overall amplitude is where you want it in your mix.

A solution to this problem is to use compression on the track. Compression helps reduce the loudness in peak areas without affecting the average loudness. Using compression in combination with an appropriate amount of makeup gain can help you increase the level on a track without obliterating the other tracks in your mix.

 For details on using EQ and dynamics processing, see the associated discussions in Chapter 8 of this book.

How to Set Levels

When it comes to setting the relative levels of your tracks, it helps to get familiar with the audio characteristics of each track first. Then consider how those characteristics contribute to the overall mix. Use the following steps as a guide to get started:

- **Listen to the track in isolation**—Use the Solo function to isolate each track in turn and familiarize yourself with its sonic characteristics. When working on a music mix, consider how the track supports the rhythm, groove, harmony, and melody of the piece. When working with nonmusical material, consider how the track contributes to the tone and emotion of the mix, and listen for the clarity of the track.

- **Listen to the track in context**—Un-solo the track to determine the appropriate level for the track relative to the other tracks in the mix. Can you still hear the important sonic elements that you heard when the track was soloed? Or is the track completely lost among competing sounds?

- **Listen to the mix without the track**—Mute the track to gauge its contribution to the mix and how necessary it is. Does the mix sound empty without the track? Or is the mix suddenly clearer without the track competing for sonic space?

- **Check the track throughout the piece**—Toggle the mute and solo state of the track on/off to hear the track isolated, in context, and removed from the mix in multiple locations throughout the composition. Are there some parts of the mix where the track is more important than others? Should the levels change for different sections of the mix?

- **Make adjustments in iterations**—Return to the track and adjust the levels as you make other changes to the mix. Setting levels is an iterative process that will require fine-tuning as your mix begins to take shape. Be sure to consider how each track's contribution changes as you begin setting levels on other tracks, adjusting pan positions, and adding processing to the mix.

Work your way through each track in the project, listening and adjusting levels as you go to create a good overall balance. Don't worry if there are aspects of the mix you aren't completely happy with at this point; you will be refining the result as you progress through later stages of the mixing process.

Panning

The pan controls on each track allow you to position the track within the stereo field. In a stereo mix, you will have two track formats to consider: mono tracks and stereo tracks.

The track format for an audio track in Reason is determined by the types of audio clips on the track and the Audio Input setting. A track with only mono audio clips, with a Mono Input, will appear as a mono track in the Main Mixer, while a track with at least one stereo clip or a Stereo Input will appear as a stereo track.

Similarly, the mono or stereo format for an instrument track is determined by the way the instrument device is connected to its corresponding Mix Channel device.

Pan Controls on Mono Tracks

Mono tracks will have a single pan knob control that allows you to position the track's output as desired in the stereo spectrum. Panning a track hard left will cause it to play out of the left speaker only; panning to center will cause the track to play at equal volume out of both speakers; and panning hard right will cause the track to play out of the right speaker only.

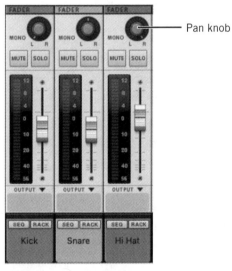 — Pan knob

Figure 7.4 Mono tracks panned hard left, center, and hard right (from left to right)

Pan Controls on Stereo Tracks

Stereo tracks will have two controls that affect panning: a pan knob control and a width knob control. By default, the width control is set to its maximum value. Lowering the width will collapse the stereo spread, reducing the panoramic width of the track. Fully lowering the width control will create a mono signal, allowing you to position the signal at a specific point within the stereo field, just as with a mono track.

On stereo tracks, you will typically leave the width control unchanged to preserve the stereo separation between the left and right channels in the source audio. You can then use the pan knob control to adjust the stereo image of the tracks.

For a stereo track where the width control is set to the maximum value, the Reason mixer performs balance panning. Turning the pan knob to the left simply turns down the right channel of the stereo signal, eventually playing only the left channel when the pan knob is set fully to the left. Likewise, turning the pan knob to the right turns down the left channel of the stereo signal. In other words, the channels of the stereo track will not appear to move across your speakers or headphones. However, if you reduce the width control to narrow the signal and then use the pan knob, you will hear Reason move the left and right channel content across the stereo image.

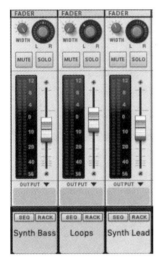

Figure 7.5 Stereo tracks with default panning (Synth Bass), mono panning (Loops), and reduced-width stereo panning (Synth Lead)

Panning Examples

As you refine your mix, you'll want to consider the panoramic position of each track, along with the overall distribution of the tracks. Positioning individual tracks at distinct locations will give each track its own space in the mix and will help you give your mix width and realism.

- **Center-panned tracks**—Certain tracks tend to work best in the center of a mix. These typically include the tracks that provide the main focus or lead part for the piece, such as a narration or voiceover track or a lead vocal track. In music production it is also common to place the main rhythmic elements, such as the kick drum, snare drum, and bass guitar, at or near the center of the mix.

- **Hard-panned tracks**—Tracks that provide flavor, color, or character to a mix are often effective when panned hard left or right. These are typically supplemental tracks, such as a shaker or harmonica in a music mix, or ambient accents, such as a distant siren or muffled argument from an adjacent room in a fiction narrative.

 Hard panning can also be effective on main tracks, especially in a sparse mix.

One potential problem with using hard-panned tracks is that they can become inaudible to a listener who is on the opposite side of the listening space from the sound source, such as when the stereo speakers are placed far apart in a living room. Also, if the stereo playback should drop a channel, everything hard-panned to the dropped side would be lost.

These problems can be mitigated by not panning fully left/right, by including some reverb from a hard-panned track in the opposite channel, and by using psychoacoustic processing to place an image in a stereo field by means of timing offsets while keeping the signal in both channels.

 For an example of hard panning, check out the electric guitar on early albums by Van Halen. Hard panning is prominent in songs such as "Little Dreamer," "Beautiful Girls," "Hang 'Em High," and numerous others.

■ **Tracks panned off center**—Often tracks are placed off center to create realism and a sense of space in the mix. This technique is commonly used to position characters around a room in a dialog mix. In music mixing, it is common to pan individual drum tracks to emulate the layout of the drum kit (placing hi-hat and splash cymbals on the left, rack toms across the middle, and the floor tom and ride cymbal on the right, for example). Another common technique is to pan backing vocal tracks, placing high harmonies on the left, midrange parts near the center, and low harmonies on the right, for example.

■ **Other ideas**—It can be very effective to offset parts that have similar impact in a mix. For example, panning a doubled vocal part to left and right extremes can create a sense of width in the mix. Similarly, panning a rhythm guitar part opposite a Rhodes piano part can help each part contribute equally without competing with one another.

 For examples of effective panning for multiple vocal parts, take a listen to "Ziggy Stardust," "Changes," and other classics by David Bowie. Bowie often used opposite-panned vocal doubles and harmonies during key sections for emphasis.

Processing Options and Techniques

When it comes to adding processing to your tracks, you have two broad categories to consider: gain-based processing and time-based processing. You also need to consider when and how any processing should be added to your mix. Here we'll cover some options available in Reason Intro. All of these options apply equally to other versions of Reason software.

Gain-Based Processing

Gain-based processors include any processors that affect the amplitude of the audio signal in some way. Examples include EQs, compressors, noise gates, expanders, and similar dynamics processors.

When adding gain-based processing to a signal, you will typically assign the processor as an insert to an individual track's mixer channel. This type of processing is usually applied to the entire signal (100 percent wet), rather than being mixed together with a dry signal. The processing is also commonly track-specific.

For example, when using an EQ plug-in to eliminate low-frequency rumble, you will want 100 percent of the source signal to be affected. And you will adjust the EQ parameters to address the specific frequency characteristics of the source signal on the track.

Time-Based Processing and Effects

The other type of processing can broadly be classified as time-based processing and effects. This includes processors that affect the signal in the time domain, such as reverbs and delays, and modulation effects, such as choruses, flangers, and phasers. These processors typically get applied to only a portion of the signal and are mixed back in with the dry signal.

Time-based processors are commonly used as shared effects, across multiple tracks in a mix. This helps provide consistency in the mix and makes more efficient use of your processing resources.

Inserts Versus Sends

Reason allows you to add signal processing to a track using inserts and sends. An insert is an audio patch point that places a signal processor directly into the signal path of the mixer channel. When using an insert, all of the audio on the mixer channel must pass through the insert on its way to the channel output.

Reason provides an insert effects section for Audio Track and Mix Channel devices, allowing you to process the track's signal through multiple successive effects devices, as needed. Reason Intro doesn't enforce a limit on the number of insert effects devices you can include, so you will instead be limited by the capabilities of your computer.

 See Chapter 8 in this book for details on using inserts for audio processing.

By contrast, a send provides a signal path that can be used to route audio from one or more source mixer channels to a parallel destination for processing. In Reason, signals from sends are returned to the mix by way of FX Return connections on the Master Section of the main mixer.

Reason doesn't use separate tracks or mixer channels for the return audio signals as is the case in many other DAWs. The return signals are controlled from the Master Section of the main mixer.

When creating a send for internal effects processing, you will start by adding a Send FX device to the rack. Then you will use the send controls on the source tracks to enable the associated send position and set the send levels from each of the tracks. The send signals route to the FX Send section of the Master Section. From there, the signal routes through the Send FX device and returns to the associated FX Return connection on the Master Section. The FX Return controls introduce the processed signal back into the mix.

Reason Intro supports eight send and return connections to the main mixer. These can be shared by any of the individual tracks in the project.

Processing with External Gear

In Reason Intro, you can also choose to route the signal from a send through external hardware, provided you have sufficient I/O on your audio interface. In this case, the send will route out of your audio interface, through an external processor, and back into your audio interface before returning to the mix.

Figure 7.6 External reverb connected to an audio interface for use with Reason Intro

Although the returning signal can be routed to FX Return connections in the Master Section for monitoring purposes, the effect will not be included if you export the song or bounce mixer channels. To hear the effect when the song is rendered, the processed signal must be recorded to an audio track. Once you've routed the return signal to an audio track, you can monitor the track input to hear the effect as you arrange and mix the song. Then you can record the processed signal when finished, prior to exporting the final mix.

Mixing in the Box

Although it is possible to use external gear for send-and-return processing in Reason Intro, mixing completely in the box offers numerous advantages. Mixing in the box simply means that all signal processing is provided by internal software devices, so the mix does not rely on any external gear.

Advantages of In-the-Box Mixing

The advantages of working completely in the box include the following considerations:

- **Portability**—Creating your Reason mix entirely in the box means that you can open the song from any Reason workstation anywhere in the world, as long as you have the song file. Barring any missing plug-ins, the mix will sound the same from any system.

  If you were to use external gear, you would need to take the gear with you to work on the mix from a different location. Or, you would need to first record the output of the external gear so that it would become audio data in the song file.

- **Recallability**—Saving the project will save all device settings and levels, allowing them to recall exactly as they were the next time the session is used. This is not the case when using external gear, as you need to reconfigure all settings on the external hardware to match your last used settings for the song.

- **Time savings**—Reconfiguring hardware is not only inconvenient; it can be quite time-consuming. It also requires keeping detailed notes of settings and configurations and updating those notes every time you change a setting.

- **Dynamic automation**—Software device settings can be automated, enabling you to incorporate dynamic changes in your mix. All automation settings are saved and recalled with the song, ensuring that it plays back consistently every time.

Getting the Most Out of an In-the-Box Mix

To get the most out of an in-the-box mix, you will want to consider how much control you need to have over the sonic characteristics of your mix. Ideally, this is something you would begin thinking about at the project inception, prior to starting to record.

For example, you'll need to decide early on if you want to be able to control the dynamics of each track entirely from within the project, or if it is okay to apply some compression to individual signals prior to the recording input stage. Likewise, you'll need to decide whether to record sources such as guitar tracks from an amp, with the guitarist's effects processing chain already in place, or to record a direct signal from the guitar and add amp simulators and stomp-box effects as effects devices inside of Reason Intro.

Aside from considerations about when to apply processing, you will also need to take steps to maximize the results of any processing you apply from within Reason Intro. Use the following simple suggestions as a starting point:

- **Learn how to set device parameters**—To effectively use the devices that are provided with Reason Intro, you will need to know how to set the parameters properly and recognize what each parameter is used for. Take time to experiment and practice using each device on material you are familiar with.

- **Learn how to use the main mixer's EQ and dynamics**—The Reason Intro mixer includes great-sounding EQ and dynamics effects built into every mixer channel. These effects are modeled from a well-known hardware mixing console. In many cases you may be able to do everything you need on a track with these tools alone, without requiring additional insert effects.

- **Avoid using copies of the same device on many different tracks**—Learn how to use send-and-return processing to apply the same effects to multiple tracks, rather than placing individual copies of the same device on each track. Not only does using multiple copies of a device require more processing power than using a single copy, but you'll find that keeping the settings in sync across multiple tracks can become tedious and time-consuming.

- **Invest in good-quality plug-ins**—The device collection that comes with Reason Intro is sufficient to get you started. As you expand beyond that basic set, invest in quality processors that enhance your sonic palette. Don't be afraid to spend some money to get what you want, but keep in mind that many high-quality plug-ins are available at very reasonable prices. Spend some time reading or viewing product reviews, and check out trial versions, if available. Once you know what you want, keep an eye out for discounts and sale prices.

(i) Plug-in manufacturers commonly offer discounts and promotional pricing during national holidays, trade shows, and similar events. If you are a student at an educational institution, many manufacturers also provide student discounts with a valid student ID.

Review/Discussion Questions

1. What are some reasons for lowering the faders across your entire project as you begin to set levels for your mix? (See "Level Considerations" beginning on page 180.)

2. Why is it important to avoid clipping in your mix? How can you tell if the project is clipping at the outputs? (See "Beware of Clipping" beginning on page 181 and "Metering in the Master Section" beginning on page 182.)

3. What are some of the meter modes provided by the Master Fader meter in Reason Intro? What are the differences between them? (See "Metering in the Master Section" beginning on page 182.)

4. How can equalization be used to help a track cut through a mix? (See "The Role of EQ and Dynamics Processing" beginning on page 184.)

5. What is meant by the dynamic response of a track? What can be done to tame excessively loud peaks on a track? (See "The Role of EQ and Dynamics Processing" beginning on page 184.)

6. How are the pan controls different between mono tracks and stereo tracks in Reason? What is the purpose of the width control? (See "Panning" beginning on page 186.)

7. Describe some examples of panning techniques that can be useful or effective in a mix. (See "Panning Examples" beginning on page 187.)

8. How does gain-based processing affect an audio signal? Give some examples of gain-based processors. (See "Gain-Based Processing" beginning on page 189.)

9. What are some available time-based processors? How are time-based processors typically applied to tracks? How is this different from the way gain-based processors are used? (See "Time-Based Processing and Effects" beginning on page 189.)

10. How are inserts different from sends in the way they process audio? What is the role of the main mixer's Master Section in a send-and-return configuration in Reason? (See "Inserts Versus Sends" beginning on page 189.)

11. What is meant by in-the-box mixing? What are some advantages of mixing in the box? (See "Mixing in the Box" beginning on page 190.)

12. Why is it important to begin thinking about in-the-box mixing from the project inception? (See "Getting the Most Out of an In-the-Box Mix" beginning on page 191.)

13. How can you maximize the results of any processing you apply from within Reason Intro? (See "Getting the Most Out of an In-the-Box Mix" beginning on page 191.)

To review additional material from this chapter and prepare for certification, see the Reason Audio Production Basics Study Guide module available through the Elements|ED online learning platform at ElementsED.com.

Creating a Basic Mix

🎧 Activity

In this exercise, you will perform basic mixing tasks in your Reason software. You'll start off by adjusting levels to get a good basic mix. Then you'll explore some strategies to prevent clipping on loud tracks. Finally, you'll adjust pan settings to enhance the mix and create space for the different instruments.

🕐 Duration

This exercise should take approximately 20 minutes to complete.

✦ Goals/Targets

- Configure the Master Fader for the project
- Set levels for different tracks in the project
- Set panning for different tracks in the project
- Save your work for use in Exercise 8

Exercise Media

This exercise uses media files taken from the song, "Lights," provided courtesy of Bay Area band Fotograf.

Written by: Zack Vieira and Eric Kuehnl; Performed by: Fotograf

The media provided for this course may be used for educational purposes only. No rights are granted to use the media for any other personal, commercial, or non-commercial purposes.

Getting Started

To get started, you will open your completed project from Exercise 6. This will serve as the starting point for this exercise.

Open your existing Lights project:

1.   Select **FILE > OPEN** or press **COMMAND+O** (Mac) or **CTRL+O** (Windows). The Browser Panel will appear (if not already visible).

2.   Use the Browser to navigate to the folder containing your completed project from Exercise 6.

3.   Locate your Lights-xxx song in the Browser.

4.   Select the song file and click **OPEN** in the Browser or double-click the song file in the Browser. The project will open as it was when last saved.

5.   If needed, choose **WINDOW > VIEW MAIN MIXER** or press **F5** to display the Main Mixer view and make it active.

Creating a Rough Mix

At this point, you're ready to start mixing. As you adjust the Volume faders on the tracks in your project, you will want to keep an eye on your output levels. However, you won't be able to use the meters on the source tracks for this since the track meters in Reason don't indicate the peak level of the full mix.

To monitor your output levels, you'll need to use the Master Fader in the main mixer.

Configure the Master Fader:

1.   Locate the Master Section on the right side of the Main Mixer view.

2.   Set the meter mode for the Master Fader by clicking its **MODE** button.

3.   Click the button repeatedly to cycle the available meter modes. Cycle modes as needed to activate **Peak** mode.

(i)   **Use the tooltip for the button to determine the current mode.**

Figure 7.7 The Peak meter mode configured for the Master Fader

4. Press the **SPACEBAR** to begin playback.

5. Listen to the levels of each track. Begin evaluating the changes you'll need to make to even out the sounds so that each part can be heard clearly without jumping out of the mix and sounding too prominent.

6. See the suggestions below to help you evaluate the tracks and begin making changes.

General Mixing Suggestions

Here are some general tips for setting the rough levels for tracks:

- Solo a track and listen closely to the content. What are the most important details of that instrument? When you un-solo the track can you still hear those details clearly in the overall mix? If not, you may need to increase the level of the track.

- Mute a track and listen to the mix. Does the mix sound empty without that track? Do you even notice that it is missing? You may decide to leave the track muted or reduce the level of the track.

- While listening through the whole song, consider whether a track is only necessary in a particular section of the song. Most songs build in complexity as they progress. You might want to consider muting or reducing the volume of a track at the beginning of the song or during verses. Then you can unmute or increase the volume of the track later in the song or during choruses.

- Keep revisiting the various sections of the song. As you make changes in one section, it may change how you feel about other sections.

Suggestions for This Project

Here are some suggestions to help you create a rough mix of your project.

- Listen to the mix going into Chorus 1, at Bar 49. The **10 Synth Lead** and **Piano** tracks are too loud in these sections. Try reducing the level of each track by about –6 dB.

- Listen to the mix in Chorus 2, at Bar 77. The **Strings** track is much too loud. Try reducing the level on this track by around –8 dB.

- Throughout the song, the vocal track (**06 Vox**) gets a little bit lost in the mix. Try increasing the level on this track by around +1.5 dB.

- The bass tracks (**05 Synth Bass** and **08 Bass Gtr**) are a bit too loud throughout the song. Try reducing the volume on each track by about –2 dB.

Adjusting Levels to Prevent Clipping

When mixing a project, it is fairly common to have clipping occur on the Master Fader at some point in a song. This can be tricky to deal with when you're first developing your mixing skills. Fortunately, you have a number of ways to address clipping in Reason.

General Level Suggestions

Here are some tips for dealing with clipping:

- If an individual track is causing clipping at the outputs (as displayed on the Master Fader meter), the first thing to do is reduce that track's level. If you can prevent clipping while the resulting level still sounds good in the mix, you're all set. If the resulting level is too quiet, you'll need to try another approach.

- Another option is to reduce the levels of all of the tracks. To do this, first reduce the level of the track that is causing the clipping until the clip goes away. Take note of how much you reduced the track from its prior level in the mix (such as –6 dB). Then reduce the level of each other track in the song by the same amount. Afterwards, you may need to bring up the level of your headphones or speaker output to hear a reasonably loud mix again.

- A third option to prevent clipping is to simply lower the Master Fader to correct the clip at the output stage. While a high level on an individual track may not be detrimental to the final mix, you never want to have clipping on the Master Fader.

Suggestions for This Project

The individual tracks in this mix are set to reasonable levels, so you should not need to reduce them further. But the overall mix is clipping. You'll use the Master Fader to solve that problem.

Correct clipping at the output stage:

1. Play through the song.

2. Keep an eye on the Master Fader, watching the level on the meter and the clip indicator at the top of the meter.

3. Take note of the loudest areas of the project. After playing the song all the way through, you should have an idea of where the signal is peaking.

4. Reduce the level of the **Master Fader** by about –4 dB to prevent clipping for this song. (Alternatively, if the Master Meter peak appears to be consistently below 0 dBFS, you could raise the level until the peak is just below clipping.)

(i) Hold Shift when moving a Volume fader to have fine control over the level.

5. Reset the Master Fader clip indicators by clicking the **RESET** button above the Master Meter.

6. Play through the song again to verify that the peak levels now remain below clipping. If clipping still occurs, reduce the Master Fader level further and test again.

Adjusting Panning

In this part of the exercise, you'll adjust the panning for your tracks. The project includes both mono and stereo tracks. Mono tracks can be easily positioned to any location in the stereo field. Stereo tracks are often adjusted to have a narrower or broader width. Both of these techniques can be used to add interest to the mix and can also create space where each track can be heard and appreciated.

General Panning Suggestions

Here are some general tips for panning tracks:

■ For each mono track, solo the track and adjust the panning. Listen for a particular pan position that feels right for the track or instrument. Un-solo the track to check how its position works in the overall mix.

■ For each stereo track, try adjusting the width control so that the width is narrower. You can often create more space in the mix by narrowing some stereo tracks. See if any tracks sound better left at the default (widest) setting instead.

■ Try panning some tracks in unusual ways. Mono tracks can be panned all the way left or right to create a novel sound. Stereo tracks can be shifted or balanced so that they are not symmetrical in the left and right channels. Just be careful because extreme panning can also distract from the listening experience.

Suggestions for This Project

Here are a few thoughts for altering the panning of your project:

- Try setting the 03 Hi Hat track slightly to the left (–15) and the 04 Claps track slightly to the right (15). This can help to create a more interesting mix of the percussion elements. (You can hear the effect of these changes in each of the choruses, where both tracks are contributing to the mix.)

- Create some space for the vocal by narrowing the width on the 06 Vox track a bit. Try setting the width control to 68.

- Narrow the stereo image of the bass guitar to help it stand out. Try setting the width knob on the 08 Bass Gtr track to 24. (You can hear the contribution of this track during each of the choruses.)

- Move the part on the Strings track to the left a bit by setting the pan control to –52. (You can hear the contribution of this track in Chorus 2 at Bar 77.)

Finishing Up

To complete this exercise, you will need to save your work and close the project. You will be reusing this project in Exercise 8, so it's important to save the work you've done.

Before you wrap up, you can also listen to the project to hear the mixing changes you've made.

Finish your work:

1. Choose FILE > SAVE to save the project.

2. Press [.] on the numeric keypad to move the Song Position Pointer to the beginning of the project.

3. (Optional) Press the SPACEBAR to begin playback and listen through the project. Press the SPACEBAR a second time when finished to stop playback.

4. Choose FILE > CLOSE to close the song.

That completes this exercise.

Chapter 8

Signal Processing

...What You Can Do to Optimize Your Audio...

This chapter begins with an overview of using effects devices and plug-ins in Reason. We take a look at some general categories of devices—including EQ, dynamics, and time-based effects—and provide some suggestions for using various processors and parameters. We also look at the process of creating send-and-return configurations and discuss how this setup can be useful.

⊕ Learning Targets for This Chapter

- Understand basic effect device functionality in Reason

- Become familiar with the major effect device categories

- Learn how to use basic device parameters

- Explore send-and-return configurations

 Key topics from this chapter are illustrated in the Reason Audio Production Basics Study Guide module available through the Elements|ED online learning platform. Sign up at ElementsED.com.

Devices and plug-ins are the key to one of the most important aspects of working with a DAW: processing audio and adding effects. Innovations in plug-in design were a key factor in the rise of the DAW as the dominant platform for producing music and creating audio for film, TV, and video games. Without great-sounding devices for equalization, dynamics processing, and time-based effects such as reverb and delay, the DAW revolution might never have occurred.

In this chapter, we take a look at how to use effects devices in Reason and review the most common types of effects devices that are used in today's production workflow.

Device Basics

The sheer variety of devices and plug-ins available today can seem completely overwhelming when you're just getting started. To ease the learning curve, Reason Studios has selected a core set of devices and included them in Reason Intro. While the selection is somewhat limited, this limitation may actually be a blessing in disguise! Learning just a few devices well is a realistic goal for those who are new to the DAW world. And, because there are only a few major categories of devices, the knowledge that you gain now can easily be applied to other devices down the road as you expand your processing palette with the Standard or Suite edition of Reason.

Viewing Inserts on Audio Track and Mix Channel Devices

Before you can place an effect device on a channel, you'll need to know how to display the Insert FX container for both Audio Track and Mix Channel devices in the Racks view.

To show the Insert FX container for an Audio Track or Mix Channel device in the Racks view:

1. Select **WINDOW > VIEW RACKS** or press **F6** to access the Racks view if it is not already visible.

2. Locate the Audio Track device or Mix Chanel device to which you would like to add an effect. If the device is currently *folded* (collapsed), click the **FOLD/UNFOLD** button (triangle) to unfold it.

Fold/Unfold button

Figure 8.1 A folded Audio Track device and the Fold/Unfold button

3. Next, click on the **SHOW INSERT FX** button to reveal the Insert FX container for the Audio Track device or Mix Channel device. (See Figure 8.2.)

Show Insert FX button

Insert FX
container area

Figure 8.2 An Audio Track device with the Insert FX container displayed

(i)   To hide the Insert FX container, click the Show Insert FX button again.

To access the Insert FX container while working in the Main Mixer view, do the following:

1. Identify the Audio Track or Mix Channel device for which you want to access the Insert FX container. Locate the corresponding channel strip in the mixer.

2. Go to the Inserts section of the mixer channel strip. Depending on the size of your window, you may need to scroll up or down in the mixer to find the Inserts controls.

3. Click the **EDIT INSERTS** button. Reason will show the Racks view (if not already visible) and will jump to the corresponding Audio Track device or Mix Channel device. The Insert FX container for the device will automatically display.

Edit Inserts button

Figure 8.3 The Inserts section of a mixer channel strip and the Edit Inserts button

Inserting Effects on Audio Track and Mix Channel Devices

Once you've displayed the Insert FX container for a device, you can insert effects devices for the corresponding channels. You'll then be able to use the devices to process the audio in various ways.

To insert an effect device:

1. Open the Insert FX container for an Audio Track or Mix Channel device as described above.

2. Choose **WINDOW > SHOW BROWSER** or press **F3** to display the Browser, if not already visible.

3. Click on the **EFFECTS** device location in the Locations List.

4. Click and drag an effect device from the Browser List to the Insert FX container of the target Audio Track or Mix Channel device. Reason will display an orange divider with a + symbol indicating the location at which the effect will be added.

5. Release the mouse button to add the device at the indicated location. Reason will automatically make cabling connections on the back of the rack to connect the device into the audio signal flow.

Figure 8.4 Dragging a Scream 4 Distortion device to the Insert FX container of an Audio Track device

6. Repeat Steps 4 and 5 to add more effects devices as needed.

When you add more than one effect device to an Insert FX container, Reason connects the devices to form a sequential signal chain, with audio passing from one device to the next. Each device in the chain adds its effect to the audio from the previous device. The devices are therefore said to be connected *in series*.

As you create effect devices in an Insert FX container, the audio signal flow follows the visual order of the devices from the top to the bottom of the screen. The devices at the top of the Insert FX container will process the audio first, while devices at the bottom of the container will process the audio last.

Processing order can be very important when setting up multiple insert effects. Putting a compressor device before a reverb device, for example, could produce very different-sounding results than putting the compressor after the reverb. Therefore, when dragging an effect device to an Insert FX container, it is important to consider whether you want to drop the new effect above, below, or between existing devices. (See Figure 8.5.)

 For more information about changing signal flow as you move existing effects devices, see "Moving and Duplicating Effects Devices" later in this chapter.

 If you use the back view of the rack to manually repatch the cables connecting devices, the signal flow may no longer follow the visual top-to-bottom order of devices.

Figure 8.5 Adding an MClass Equalizer device between two other effects devices on an audio track

To replace an existing insert effect device with a different type of device:

1. Drag a device from the Browser as though you were going to add it to the Insert FX container.

2. Drop the new device directly on top of an existing device. Reason will show an orange overlay with a "replace" icon over the device that will be replaced. (See Figure 8.6.)

Figure 8.6 Replacing a Scream 4 Distortion device with an Audiomatic Retro Transformer device

To remove an insert effect device:

1. Click on the device in the Racks view to select it.

2. Do one of the following:

    • Press the **DELETE** or **BACKSPACE** key.

    • Choose **EDIT > DELETE DEVICES AND TRACKS**.

    • Right-click on the background area of the device outside of any of the controls and parameters. Choose **DELETE DEVICES AND TRACKS** from the pop-up menu.

3. Click the **DELETE** button in the confirmation prompt to verify that you want to delete the device.

> (i) To delete a device without receiving a warning prompt, press Command+Delete (Mac) or Ctrl+Delete (Windows).

If you have multiple effect devices chained together and you delete one of the devices, Reason will automatically update the cable connections for the remaining devices on the back of the rack to keep the signal flow intact. (See Figures 8.7 and 8.8.)

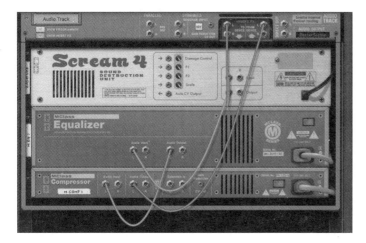

Figure 8.7 A Scream 4 device, MClass Equalizer device, and MClass Compressor device are connected in series as insert effects.

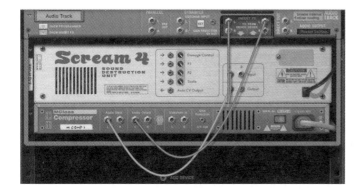

Figure 8.8 After deleting the MClass Equalizer device, the Scream 4 device is now connected to the MClass Compressor device.

To remove all insert effects from an Audio Track or Mix Channel at once, do the following:

1. Select the target device:

    • In the Racks view, click on an Audio Track or Mix Channel device to select it.

    • In the Main Mixer view, click on a channel strip corresponding to an Audio Track or Mix Channel device to select it.

2. Choose **EDIT > CLEAR INSERT FX**. All Insert FX will be removed from the associated channel.

Moving and Duplicating Effects Devices

After you've begun working with effects on tracks in a song, you may find that you need to move an effect device to a different order in the signal flow or to a different track entirely. You may also want to create a duplicate of the effect elsewhere in the project. These tasks are easily accomplished in Reason.

To move and reroute an effect device, do the following:

1.   Click on the device.

2.   While holding the **SHIFT** modifier, drag the device to a new position in the rack.

     You can move the device to a different position in the current Insert FX container, or you can move the device to the Insert FX container of a different Audio Track or Mix Channel.

Holding the **SHIFT** modifier while moving an effect device is important. If you do not hold **SHIFT** while moving a device in the rack, the device will visually move to the new location, but the cable connections on the back of the device will not be updated. As a result, the actual audio signal flow will not change.

The **SHIFT** modifier tells Reason to re-cable the device and insert it into the signal flow at its new location, in the same way that a newly created effect device would be routed at that location. If the signal flow is not updated, you may later realize that you aren't hearing the results you expect. At that point you might have to look at the back of the rack and check the cable connections to determine what went wrong.

 If you want to visually change the location of a device without rerouting it, just omit the Shift modifier while moving the device.

To duplicate a device and automatically connect the new copy into the signal flow:

1.   Click on the original device to select it.

2.   Do one of the following:

     • Hold **SHIFT** and choose **EDIT > DUPLICATE DEVICES AND TRACKS**.

     • Right-click on the background area of the device outside of any of the controls and parameters. Hold **SHIFT** and choose **DUPLICATE DEVICES AND TRACKS** from the pop-up menu.

     • **SHIFT-OPTION-CLICK** (Mac) or **SHIFT-CTRL-CLICK** (Windows) on the background area of the device outside of any of the controls and parameters and drag the device to the desired location.

 On Windows, you will need to press the CTRL modifier after you start dragging the device; if you hold CTRL when you initially click on the device, you will instead open the pop-up context menu for the device.

As when moving a device, holding the **SHIFT** modifier in all of the actions above will cause Reason to automatically connect the duplicated device into the signal flow. If you do not use the **SHIFT** modifier, the new device will not be connected to anything. You would then need to manually connect the device using the back view of the rack.

You can also use a keyboard shortcut to activate the **EDIT > DUPLICATE DEVICES AND TRACKS** menu command: **COMMAND+D** (Mac) or **CTRL+D** (Windows). However, you can't use the **SHIFT** modifier with these commands, so duplicating a device with the menu keyboard commands will always create a disconnected device.

Displaying VST Plug-In Windows

For devices built into your Reason software and Rack Extension plug-ins, you can adjust device parameters directly in the Racks view. However, VST plug-ins function slightly differently. VST plug-ins are also represented by devices in the rack, but to adjust the parameters of a VST plug-in, you'll need to open the plug-in window.

To open a VST plug-in window:

1. Locate the VST plug-in rack device in the Racks view.

2. If the device is unfolded, click the **OPEN** button or the display area showing the words VST Plugin. (See Figure 8.9.)

(i) If an image of the VST plug-in has been saved, it will appear in place of the words VST Plugin.

If the device is folded, click the smaller **OPEN** button that appears near the right side of the device.

Figure 8.9 An unfolded VST plug-in device inserted on an Audio Track device

The VST plug-in window will open, showing the available plug-in parameters. (See Figure 8.10.)

Figure 8.10 An open VST plug-in window appearing near the VST plug-in rack device

Working with Devices

Interacting with devices in the Racks view is a common task in Reason. The software provides many tools and techniques for managing devices that can be helpful as you venture beyond the basics.

Shortcuts for Adding Insert Effects

Working with insert effects by going directly to the Insert FX container of an Audio Track or Mix Channel device is an important skill; you need to understand how to work with the container in order to manage your effects devices. However, Reason provides some additional time-saving options for adding insert effects to a track.

Rather than first opening the Insert FX container and dragging an effect to it, you can create an effect directly on a selected track.

To create an effect on a selected track, do the following:

1. Click on any one of the following items to select it:

    • An Audio Track device in the Racks view

    • A Mix Channel device in the Racks view

    • A channel strip in the Main Mixer view

    • An audio track in the Sequencer view

    • A mix channel automation track in the Sequencer view

A Mix Channel track is not the same as an instrument track. Selecting an instrument track will add the effect directly to the associated instrument, as discussed later in this chapter.

2. Access the Browser and locate the effect device you would like to add.

3. Do one of the following:

    •   Double-click the desired effect device in the Browser.

    •   Select the desired effect device in the Browser and click the **CREATE** button at the bottom of the Browser.

Alternatively, if you are already working in the Sequencer view, you can drag and drop devices without switching to the Racks view.

To add effects in the Sequencer view, do the following:

1. Access the Browser and locate the effect device you would like to add.

2. Drag the effect device from the Browser to the Track List of the Sequencer view.

3. Position your mouse cursor beneath the track to which you would like to add the effect. Reason will display an orange line with a + symbol below the track where the device will be added.

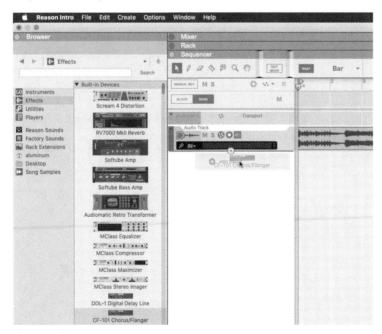

Figure 8.11 Dragging an effect device to an audio track in the sequencer

4. Release the mouse button to drop the device and add it to your song.

    The exact location where the device is added will depend on the type of track. For audio tracks and mix channel automation tracks, the effect will be added to the Insert FX container. For instrument tracks, the effect will be connected directly to the instrument device.

When you add an effect using these techniques, Reason will show the Racks view, if it is not already visible. The Insert FX container will automatically display and scroll the new device into view.

Effects Devices and Instrument Devices

Instrument devices provide another possibility for working with effects. Here is a quick recap of instrument functionality discussed in Chapter 6:

■   An instrument track contains note data that is sent to an instrument device.

■   An instrument device converts note information into an audio signal that can be heard.

■   Each instrument device is cabled to a Mix Channel device in the rack, causing the audio signal to flow from the instrument to the Mix Channel device.

■   An instrument's Mix Channel device passes the instrument's audio signal through a channel strip in the main mixer.

■   Ultimately, the output of an instrument's channel strip is blended with other signals in the mixer, passed through the Master Section, and then out through your audio interface and listening device.

Up to this point, we have discussed adding insert effect devices to Insert FX container areas of either Audio Track or Mix Channel devices. For audio tracks, adding insert effects in this way is effectively the only possible approach.

For instrument devices, though, it's possible to drag and drop effects devices beneath the instrument device itself, rather than adding the effects to the Mix Channel for the instrument. In Figure 8.12, a Subtractor synthesizer is connected directly to an Audiomatic device. The signal flow then goes to a Mix Channel, which contains a different Audiomatic device.

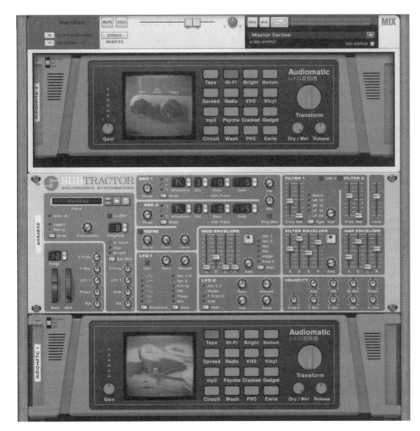

Figure 8.12 A Subtractor connected directly to an Audiomatic device, with another Audiomatic device inserted on the Mix Channel

With this kind of configuration, the audio signal will pass from the Subtractor to the lower Audiomatic device that connects directly to the Subtractor. From there, the signal will continue on to the Mix Channel device, where it will eventually reach the upper Audiomatic device inside the Insert FX container.

A natural question that arises here is, "Does it make a difference whether an effect device is placed directly on an instrument device or in the Mix Channel Insert FX container?" In most cases, the answer is no, there will not be a noticeable difference. In both cases, the effect device will process the audio coming from the instrument before it reaches the mixer output.

Where there can be a difference, however, is in the application of mixer channel strip processing, such as the built-in dynamics and EQ effects (discussed later in this chapter). By default, these effects come *before* the effects in the Insert FX container of the Mix Channel. By contrast, they are applied *after* any insert effects that are connected directly to an instrument device.

Ultimately, this lets you choose where you would like to place effects devices in the instrument signal flow. Putting the effects inside the Mix Channel Insert FX container is generally preferable, as it provides a number of benefits:

- Closing the Insert FX container hides all of the insert effect devices, reducing visual clutter in the Racks view.

- The mixer channel strip provides macro button and knob controls that you can program to change device parameters of effects within the Insert FX container.

- You can save a patch file for the Insert FX container. This will store the layout and configuration of all of the devices inside the Insert FX container. The saved patch file can be loaded on other Audio Track or Mix Channel devices to instantly re-create the same set of effects elsewhere in a song.

- You can easily delete all of the effect devices with the **CLEAR INSERT FX** command discussed earlier in this chapter.

 Channel strip macro control and insert effect patch file functionality are outside the scope of this book. Consult the operation manual for your Reason product for details on using these features.

The Disconnect Device and Auto-Route Device Commands

Reason software provides two commands that can be helpful if a device gets into a position where it does not seem to be routed correctly. The Disconnect Device command removes all cable connections from a device. The Auto-Route Device command reconfigures the cable routing on the back of a device to match the way the device would be cabled when first created.

To access the routing commands:

1. Click on the target device in the Racks view to select it.

2. Do one of the following:

   - Choose the desired menu command: **EDIT > DISCONNECT DEVICE** or **EDIT > AUTO-ROUTE DEVICE**.

   - Right-click on the background area of the device (outside of the controls and parameters) and choose the desired command from the pop-up menu: **DISCONNECT DEVICE** or **AUTO-ROUTE DEVICE**.

Choosing Disconnect Device followed by Auto-Route Device will fully reset device routing. You can always check the back view of the rack to verify the cable connections to a device.

Device Controls

Reason devices feature two sets of controls: a common group of controls that appears on most devices, and device-specific parameters that are unique to an individual device. Each VST plug-in window also has a common set of controls in a toolbar at the top of the window. We'll start by taking a look at the common controls. Then we'll look at process-specific controls for device parameters.

Common Device Controls

Controls that are commonly found on devices include the power/bypass switch and patch section controls for browsing, selecting, and saving presets. Other controls that are common for VST plug-ins include the four buttons in the VST plug-in window toolbar.

Power/Bypass Switch

On the upper left of every effect device in Reason you can find the Power/Bypass switch. This control looks like a three-way switch that can be set to Bypass, On, or Off.

Figure 8.13 The Power/Bypass switch on the left side of an MClass Compressor device

You can set the switch to a desired position by clicking and dragging on the switch control itself or by clicking in the background of the switch area to the left of the words Bypass/On/Off.

Adjusting the switch changes how the effect device operates:

- **Bypass**—In this mode, audio signals pass through the device without any changes. You can use the Bypass setting to hear the audio without the effect device applied. For example, you might toggle a compressor device back and forth between Bypass and On to compare the sound of the uncompressed audio to the compressed audio. This technique is referred to as an A/B comparison.

- **On**—In this mode, audio signals flow through the device and the effect of the device is applied. On is the default state for an effect device. For example, when a compressor device is switched on, it will process the incoming audio signal and output an audio signal with a compressed dynamic range.

- **Off**—In this mode, the audio signal flow through the device is broken. The device will perform no processing, and no signal will come out of the audio outputs for the device. Because the signal flow is stopped entirely, if the device is part of the insert signal flow for a track, the track will effectively become muted. This setting is useful primarily for devices connected in a send-and-return configuration, to temporarily disable the effect for all tracks.

Patch Section

Some effect devices (although not all of them) contain a set of common controls in an area called the Patch Section. This section does not always appear at the same location in each device, but it always contains the same elements. The Patch Section controls are used to browse, load, and save patch files for the device. Patch files are often referred to as *presets*, and they contain stored settings for the parameters of a device that can be loaded to produce a particular sound.

Figure 8.14 The Patch Section controls (as displayed on the Scream 4 Distortion device)

The Patch Section contains the following controls:

- **Patch Name display**—This text display shows the name of the currently loaded patch or the words Init Patch if a patch has not been loaded. You can click on the patch name to open a menu and select a patch to load. The menu lists patches from the same folder as the patch that is currently loaded.

- **Select Patch buttons**—The up arrow and down arrow buttons load the previous or next available patch for the device, respectively. These buttons cycle between patches in the same folder as the patch that is currently loaded.

- **Browse Patch button**—This button displays the Browser and sets the browse context for the effect device. This lets you find patch files in the Browser to load into the device. While the browse context is focused on the effect device, you can double-click a patch file in the Browser List to load it or select the patch file and click the **LOAD** button at the bottom of the Browser.

- **Save Patch button**—This button opens a file save prompt where you can enter a file name to store your own patch file that contains the current settings configured for the device.

VST Plug-In Window Controls

The top of every open VST plug-in window contains a toolbar with a common set of controls that help to manage VST plug-in devices.

Figure 8.15 The VST plug-in window controls

The following buttons comprise the plug-in VST window controls:

- **Keep Open**—Normally if you have a VST plug-in window open and then click on another device in the Racks view, Reason will automatically close the plug-in window. Click the **KEEP OPEN** button to prevent the window from closing when you select another device. Clicking the button again will turn it off and re-enable automatic closure of the plug-in window.

■ **Automate**—This button allows you to select a parameter in the VST plug-in window for automation. Click the **AUTOMATE** button in the toolbar, and then click on the desired parameter in the VST plug-in window.

 Automation is discussed in more detail in Chapter 9 of this book.

■ **Remote**—This button allows you to assign a control on your MIDI controller to a parameter of the VST plug-in. Click the **REMOTE** button and then click on the desired parameter in the VST plug-in window to begin.

 Remote Override features in Reason are not covered in this book. See the Operation Manual for your Reason product for detailed information.

■ **Screenshot**—This button captures an image of the current appearance of the VST plug-in window. The screenshot image will appear in the Browser and on the unfolded VST plug-in device in the rack in the future.

 Capturing screenshot images of your VST plug-ins will help you visually identify them in the Browser and in the rack.

Device-Specific Parameters

Once you have an effect device added to an Insert FX container with the parameters visible, you're ready to start adjusting the device's settings. This is where the fun begins! Let's take a closer look.

The most straightforward way to adjust device parameters is to use the mouse. For fader-style controls, simply click and drag up to increase the value of the control or drag down to decrease the value. To adjust knob-style device parameter controls, you will also click and drag vertically. Dragging up will increase the value of a knob-style control, whereas dragging down will decrease the value.

 Reason devices do not respond to clicking at the location where you'd like a knob to go. You must click and drag to change the knob's position.

You can change the way that device controls respond to the mouse using keyboard modifiers.

For fine adjustments:

■ Hold **SHIFT** while moving the control.

To return a control to its default value:

■ **COMMAND-CLICK** (Mac) or **CTRL-CLICK** (Windows) on the control.

EQ Processing

EQ (short for equalization) is probably the most commonly used processor in the audio universe. In the consumer electronics world, EQ can be found just about anywhere that an audio signal is present; even the humblest car stereo usually has basic low, mid, and high EQ controls.

In the realm of audio production, EQ is an absolutely essential tool for getting individual instruments to sound their best and for getting a multitrack mix to sound great.

Types of EQ

In the simplest terms, EQ is used to manipulate the frequency component of an audio signal. However, EQ processors themselves come in a variety of flavors that offer different numbers of EQ bands and different configurations within those bands.

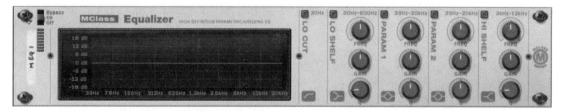

Figure 8.16 The MClass Equalizer device in Reason Intro

Some types of EQ bands include the following:

■ **Parametric**—This is the most common type of EQ band. A parametric band is fully adjustable in terms of the frequency, gain, and Q (width) parameters.

■ **High- and Low-Pass Filters**—These constitute another common type of EQ band and are often abbreviated HPF and LPF, respectively. These filters permit signal to "pass" only above (HPF) or below (LPF) the selected frequency. High-pass filters are commonly used to eliminate low-frequency rumble, while low-pass filters can be applied to remove high-frequency hiss.

■ **Shelf**—The shelf is similar to the high- and low-pass filter, but it is used to boost or cut the signal above (high-shelf) or below (low-shelf) the selected frequency by a specified amount.

■ **Notch**—A notch filter is essentially a parametric band that has the gain setting fixed at the maximum negative value. This type of filter is used to "notch out" an offending frequency to make it inaudible.

Basic EQ Parameters

You'll find a number of standard EQ parameters on most software and hardware EQ processors. These will vary somewhat based on the type of EQ band.

Common EQ parameters include the following:

- **Frequency**—The Frequency parameter determines the target frequency for an EQ band. Every type of EQ band will have a frequency parameter, although some may be fixed at a specific frequency.

- **Gain**—The Gain parameter is used to boost or cut the volume of the associated EQ band. A gain control is typically found on parametric and shelf bands.

- **Q**—This is probably the most confusing EQ parameter for those who are new to audio production. The Q control is used to adjust the shape (slope or width) of an EQ curve. It behaves somewhat differently depending on the type of EQ band being used (parametric, shelf, filter).

- **Band Enable**—The Band Enable parameter can be used to enable or disable an EQ band, thus controlling whether that particular band will have an audible effect on the signal. This can be useful for doing an A/B comparison of the signal with and without an individual EQ band.

- **Input/Output**—The Input and Output gain controls, when available, can be used to boost or cut the overall signal amplitude at the input (before the EQ has been applied) or output (after the EQ has been applied).

EQ Effects in Reason Intro

Reason Intro offers two types of EQ effects: the MClass Equalizer and the Main Mixer Channel Strip EQ. The MClass Equalizer rack device is a high-quality five-band EQ that offers two fully parametric bands, a low-shelf band, a high-shelf band, and a high-pass filter (labeled "Lo Cut" on the device) that has a fixed frequency of 30 Hz.

The Channel Strip EQ functions a bit differently in that you don't need to add it as a device to the rack. Instead, every channel strip for an Audio Track or Mix Channel device in the main mixer already has a built-in set of EQ controls that can be activated to process the audio. (See Figure 8.17.)

The Channel Strip EQ is modeled after the channel EQ of a hardware mixing console. It offers two fully parametric bands, a high-shelf band, a low-shelf band, a high-pass filter (HPF), and a low-pass filter (LPF).

The full version of Reason and Reason Suite offers a separate device called the Channel EQ that allows you to apply the Channel Strip EQ effect as an independent device in the rack, similar to the MClass Equalizer.

Figure 8.17 The Channel Strip EQ controls in the main mixer

The Channel Strip EQ contains three separate On buttons that determine which components of the EQ are active. The three On buttons correspond to the low-pass filter, the high-pass filter, and the group of four remaining EQ bands. You will need to turn on the components you would like to use; when an EQ component is turned off, its processing is bypassed, and the corresponding knobs have no effect.

The EQ controls also feature two other sets of buttons:

- **Bell**—By default, the red high-frequency (HF) band and the gray low-frequency (LF) band in the EQ are shelf bands. Each of these two bands has a Bell button that you can click to switch the band from a shelf to a bell-shaped EQ curve. Note that activating Bell mode does not make these bands fully parametric; they have a fixed Q (width) that you cannot change.

- **E**—The E button changes the behavior of the two middle, fully parametric bands of the EQ, labeled HMF and LMF. When the button is disabled, the EQ gently increases in width when the gain (dB knob) is set to low values and narrows in width when the gain is set to higher values, which can help to impart a smoother and more "musical" character to the EQ. When the button is enabled, the width does not change relative to the gain and remains fixed at the value you specify with the Q knob.

The Channel Strip EQ has another very useful ability that you won't find on the MClass Equalizer. The Spectrum EQ Window button (curved line above the LPF controls) opens a separate window that allows you to both view a real-time spectrum analysis of the audio and graphically change the EQ parameters.

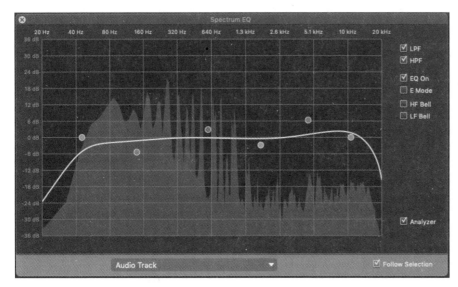

Figure 8.18 The Spectrum EQ window

The Spectrum EQ window shows the frequency information for the audio as a gray graph in the background. The foreground contains a white line showing the cumulative EQ curve produced by the current EQ settings. The colored circles represent controls for the individual EQ bands that you can click and drag. Adjusting the EQ settings in the Spectrum EQ window will also update the settings on the channel strip.

The checkboxes on the right side of the window correspond to the On, Bell, and E buttons found in the channel strip.

 You can also open the Spectrum EQ window by pressing Function Key F2 or by choosing Window > Show Spectrum EQ Window.

Strategies for Using EQ

Different types of material will require different approaches to EQ. But the basic techniques for finding good EQ settings are similar from track to track. Following are some tips for using the Channel Strip EQ and its bands.

To use the Channel Strip EQ on a channel:

■ Enable the **HPF** and increase the frequency enough to eliminate any unwanted noises at extreme low frequencies, but not so much that the signal loses too much low end.

On some tracks, such as drum overheads, you may want to eliminate almost all of the bass to make room for the kick drum track and other low-frequency instruments.

- Use the **HMF** or **LMF** band to emphasize desirable qualities of a particular channel.

  A common approach used to identify a target frequency is to boost the gain substantially (+12 dB or more) and slowly sweep the frequency control back and forth during playback. Listen carefully to the channel while adjusting the frequency. Stop sweeping when you hear something you like, tone-wise. Then reduce the gain until the frequency is only slightly pronounced, compared to the original.

- Use the **HMF** or **LMF** band to reduce undesirable qualities of a particular channel.

  A similar technique to the above can be used to identify target frequencies to cut. Simply stop sweeping when you hear something in the tone that sounds bad, intrusive, or unwanted. Then adjust the gain to a negative value that effectively reduces the prominence of the frequency compared to the original.

- If necessary, use an **HF** or **LF** shelf to add some high-end sizzle to cymbals or low-end thump to a bass guitar (respectively).

- Frequently toggle the On buttons for the EQ to make sure that the changes you're making are actually improving the sound of the channel and helping it to sit better in the overall mix!

Dynamics Processing

Like EQ, dynamics processing is an essential part of the audio production workflow. Unlike EQ, dynamics processing is not particularly well known outside of audio production. It is fair to say that working with dynamics processors is a bit more complex than using EQ to boost the bass or cut the treble on your car stereo. But the basics of dynamic processing are pretty straightforward, and if you use your ears to dial in just the right settings, you'll be working like a pro in no time!

Figure 8.19 The MClass Compressor in Reason Intro

Types of Dynamics Processors

Dynamics processing refers to the manipulation of the volume (or amplitude) of an audio signal. Many instruments benefit from dynamics processing, including vocals, guitar, bass guitar, and drums. Depending on the situation, dynamics processors can be used to control a signal level that is varying dramatically or to isolate just the loudest part of a signal while reducing or eliminating the quiet parts.

Some types of dynamics processors include the following:

- **Compressors**—The compressor is the most common type of dynamics processor. It is used to reduce the loudest part of an audio signal, resulting in a more predictable dynamic range. Once the loudest parts have been reined in, it becomes possible to boost the entire signal without fear of clipping.

- **Limiters**—A limiter is essentially a compressor with a very high ratio setting (see the "Basic Dynamics Parameters" section below). As a result, the incoming audio signal is completely capped at a specified level that prevents clipping. Limiters are an essential part of the music production workflow and are often used to make an overall mix seem louder.

- **De-Essers**—A de-esser is a compressor that operates at a specific frequency. Its primary use is to reduce excessive "s" sounds in a recording, a phenomenon known as *sibilance*. This type of processor can also be used to eliminate other undesirable high-frequency sounds, such as the breath noise in a flute recording.

- **Expanders**—An expander is a dynamics processor that can be used to reduce the level of a signal when it falls below a certain level. Expanders work something like a compressor in reverse. A gentle amount of expansion can help to reduce some of the ambient noise that gets picked up in an audio recording.

- **Gates**—A gate (often called a *noise gate*) is a type of expander that will completely silence a signal when it falls below a certain level. Gates are frequently used to eliminate the "bleed" that occurs when a microphone picks up signal from an off-mike sound source, such as a snare drum being picked up by a kick drum mike. Gates are also useful to eliminate background noise in a recording, such as the buzz of a guitar amp that can be heard in silent parts of the performance. The purpose of a gate is to clean up the sound of a recording by helping to isolate the desired parts of the audio signal and eliminate the unwanted parts.

Basic Dynamics Parameters

Just like EQ processors, dynamics processors make use of a standard set of controls. These will vary somewhat based on the type of dynamics processor being used.

Common dynamics parameters include the following:

- **Threshold**—The threshold parameter sets the level that the input signal must reach in order to trigger the processor. Note that compressors and limiters are triggered when the signal exceeds the threshold, whereas expanders and gates are triggered when the signal drops below the threshold.

- **Ratio**—The ratio parameter determines the amount of signal reduction that occurs once the threshold has been reached. For example, when using a compressor with the ratio set to 2:1, the signal must exceed the threshold by 2 dB to result in an output increase of 1 dB. A limiter typically features a compression ratio of 100:1.

- **Knee**—The knee parameter determines how suddenly or gradually gain reduction is introduced in a compressor as the signal level approaches the threshold.

- **Gain**—The output gain parameter is used to add a level increase (known as *makeup gain*) to a signal that has been reduced in volume by a dynamics processor.

- **Attack/Release**—The attack parameter determines how quickly the dynamics processor will act once the audio signal has reached the threshold. The release parameter determines how quickly the dynamics processor will disengage once the threshold is no longer met.

Dynamics Effects in Reason Intro

Reason Intro offers four types of dynamics effects: the MClass Compressor, the MClass Maximizer, the Master Compressor, and the Channel Strip Dynamics.

The MClass Compressor is a versatile compressor that features typical compression controls as well as a Soft Knee mode and an Adapt Release mode. The MClass Maximizer is a loudness-maximizing limiter; it is designed to be applied to an entire stereo mix to increase the overall volume of a song, but it can also be used as a limiter on individual tracks. The Master Compressor appears in the Master Section strip on the right side of the main mixer and is not available as a device that can be added to the rack. The Master Compressor is used to add compression to an entire mix; this can impart an extra sense of polish and "glue" to the final product.

Like the Channel Strip EQ, the Channel Strip Dynamics effect is worth a closer look. Every channel strip in the main mixer contains a dynamics section that includes compression and expansion/gating effects.

 The full versions of Reason and Reason Suite offer two separate devices: Channel Dynamics, which allows you to apply the Channel Strip Dynamics effect as an independent device in the rack, similar to the MClass Compressor, and Master Bus Compressor, which allows you to apply the Master Compressor as an independent device in the rack.

Figure 8.20 The Channel Strip Dynamics controls in the main mixer

The upper Comp (compression) section and lower Gate section each have independent On button controls that you will need to enable for the effects to be active. The dynamics effects contain many of the standard parameters described above, as well as a few additional controls that you'll need to understand:

- **Fast button**—Neither the Comp nor the Gate controls contain adjustable attack time knobs. Instead, you can use the Fast button to switch between fixed fast and slow attack times.

- **Peak button**—Found in the Comp section only, this button changes the way the compressor measures the amplitude of the audio signal against the threshold. When the Peak button is disabled, the compressor acts on the average RMS level of the audio. When Peak is enabled, the compressor acts on the peak level, resulting in a faster and more accurate response to signals with sharp peaks (such as drums).

- **Exp button and Range knob**—Found in the Gate section only, the Exp button toggles the Gate section between gating and expansion. When the button is disabled, this effect acts as a gate and reduces signals below the threshold by an amount specified with the Range knob. When the Exp button is enabled, the effect acts as an expander, and the Range knob instead sets a gain reduction amount, relative to how far below the threshold the signal drops.

- **Hold knob**—Found in the Gate section only, the Hold knob sets an additional wait time that must elapse before the effect begins acting on the audio after the signal drops below the threshold. This hold time occurs before the effect applies the release time. The Hold knob is useful to prevent the gate from opening and closing quickly when a signal stays close to the threshold.

Applying dynamics processing to tracks can seem confusing at first. But, just like when working with EQ, there are some common strategies to help you find the right settings.

Strategies for Using Compression

To add compression to a channel in Reason, you can add the MClass Compressor device to the Insert FX container of the target Audio Track or Mix Channel device.

To use an MClass Compressor device on a channel:

1. Set the **ATTACK** to about 10ms and the **RELEASE** to about 100ms.

2. Start off with a **RATIO** of about 3.0:1. (You may need a higher ratio if the signal has a wide dynamic range.)

3. Reduce the **THRESHOLD** setting until you start seeing a small amount of compression registering in the gain reduction (GR) meter, labeled **GAIN** on the device. A good starting point is to target around 3 to 6 dB of gain reduction.

4. Adjust the **OUTPUT GAIN** control to increase the overall signal level.

Make sure that you don't increase the gain so much that the song begins clipping. If you're having trouble finding the perfect gain setting, you may need to increase the ratio or lower the threshold.

 Reason Intro does not provide a dedicated, built-in de-esser device. It is possible to create the function of a de-esser using Reason Intro's built-in effects. The Factory Sounds library includes a few patches for the Combinator device that implement this type of custom-built de-esser.

Reverb and Delay Effects

Reverb and delay effects are indispensable tools for creating a sense of space. Collectively known as *time-based effects*, reverb and delay can be used to create a wide range of sounds. This range spans the gamut from providing tasteful spatialization of audio, to adding a subtle spaciousness to a track, to creating dramatic effects that completely alter the sound of a voice or instrument.

What Is Reverb?

Reverb effects are used to create the sound of a real or imaginary space. Reverb algorithms can be used to re-create the sound of small spaces (rooms, cars, phone booths) or large spaces (concert halls, parking garages). A number of types of reverbs are available in both hardware and software formats. Some use a series of delays (see "What Is Delay?" below) to create smooth, musical-sounding results. Other reverbs use a technique called *convolution*, which actually uses samples of real-world acoustic spaces to create incredible realism.

Figure 8.21   The RV7000 MkII Reverb device in Reason Intro

Reverb in Reason Intro

Reason Intro offers a sophisticated reverb device called RV7000 MkII. RV7000 MkII offers a variety of features that are an excellent introduction to the world of reverb.

To access all of the parameters of the RV7000 MkII, you may need to unfold it twice. If the main device is not unfolded, first click the Fold/Unfold button at the top left, next to the Power/Bypass switch. Then click the arrow button next to the words "Remote Programmer" to access the full set of device parameters.

Controls in the top section of the RV7000 MkII include the following:

- **Decay**—The Decay control determines how long it will take for the reverb to fade out after a sound has been processed.

- **HF Damp**—The HF Damp knob changes the rate at which the high frequencies of the reverb effect decay (fade out) relative to the other frequencies. Increasing the HF Damp control fades out the high frequencies more quickly and creates a darker reverb sound.

- **HI EQ**—The HI EQ control adjusts the gain of a high-shelf EQ band for the reverb effect. Raising the HI EQ control adds a high-frequency boost to the reverb effect.

- **Dry-Wet**—The Dry-Wet control can be used to adjust the balance of the device's output between the input signal and the reverb effect signal. (See "Wet Versus Dry Signals" below.)

Controls in the Remote Programmer section of the RV7000 MkII include the following:

- **Algorithm**—The Algorithm control offers a number of options, including Small Space, Room, Hall, Arena, Plate, Spring, Echo, Multi Tap, Reverse, and Convolution. These algorithms fundamentally change the nature of the reverb.

- **Size**—The Size control is used to choose the size in meters of the space simulated by the currently selected algorithm. Not all algorithms provide a size parameter.

- **Diffusion**—When set to a high value, the Diffusion control will emphasize the initial build-up of echoes. Low values will reduce the initial buildup, which can result in better clarity. Not all algorithms provide a diffusion parameter.

- **Room Shape**—For the reverb algorithms that simulate types of rooms (Small Space, Room, and Hall), the Room Shape parameter changes between different variations of the simulated space that provide slightly different character for the reverb effect.

- **Predelay**—The Predelay control adjusts the amount of time that the reverb waits after receiving an input signal before it begins to create reverb. Not all algorithms provide a predelay parameter.

Applications for Reverb Processors

You'll find many creative ways to use reverb in a project. Whether adding a subtle sense of space to a vocal or completely washing out a guitar to create an ambient bed, reverb can truly bring a track to life.

Here are some suggestions and scenarios for reverb usage:

- Add some character to a snare track by adding a room reverb directly as an insert effect. Then adjust the Dry/Wet control to get the proper balance.

- Give a vocal track a vintage sound by using a medium plate reverb.

- Create a distant, washed-out guitar sound by using a large hall reverb and setting the Dry/Wet control to 100% wet.

- Add a medium hall reverb in a send-and-return configuration to apply reverb to a number of tracks in a session, including vocals, acoustic guitars, keyboards, and more. This can help gel the mix by putting all of the components into the same acoustic space.

What Is Delay?

Delay is the term that audio engineers use for a processor that creates an echo. Delay is a simple effect in concept, but over the years both hardware and software delays have evolved into extremely complex processors.

Delay devices provide dozens of different approaches to delay, from digital models of vintage tape-style echoes to modern granular pitch-shifted delays. Almost all modern delays have the ability to synchronize their individual echoes to the tempo of a song.

Figure 8.22 The DDL-1 Digital Delay Line device in Reason Intro

Delay in Reason Intro

Reason Intro offers a delay device called DDL-1 Digital Delay Line. DDL-1 is a simple but powerful delay effect that can generate anything from single echoes to beat-synchronized delays. You can chain together multiple DDL-1 devices to create even more complex delay effects.

Controls in DDL-1 include the following:

- **Delay Time**—The numeric display and up/down arrows are used to set the device delay time, with a range from **1 ms** to **2,000 ms** (two seconds).

- **Unit switch**—The Unit switch changes the DDL-1 between **MS** and **Steps** mode. In **MS** mode, the delay time can be set freely in milliseconds and does not follow the song's tempo. In **Steps** mode, the delay time follows the song's tempo as determined by the Step Length switch.

- **Step Length switch**—The Step Length switch has an effect only when the Unit switch is set to Steps mode. This switch determines the unit of measure that the delay time uses, such as 1/16 note or 1/8 note triplets. For example, if the delay time is set to 2, the Unit switch is set to Steps, and the Step Length switch is set to 1/16, the delay time will be two 1/16 notes, at the song's current tempo.

- **Feedback**—The Feedback control determines how much of the device's output will be "fed back" into the input. The feedback setting ultimately controls the number of echoes that will occur.

- **Pan**—The pan control sets the position of the delay echoes in the stereo field, just like panning a channel in the main mixer.

- **Dry/Wet**—The Dry/Wet control can be used to adjust the balance of the device's output between the input signal and the delay echo signal. (See "Wet Versus Dry Signals" below.)

Applications for Delay Processors

A little delay can help add excitement to your mix. This is a great way to propel a track forward and add rhythmic variety.

Here are some suggestions and scenarios for delay usage:

- Add some width to a mono acoustic guitar track by applying a panned delay device. Or, for an advanced exercise, dive into routing in the rack and use multiple DDL-1 devices in conjunction with the Spider Audio Merger & Splitter or Line Mixer 6:2 utility devices to set up independent delays for the left and right channels with slightly different delay times.

- Apply a simple quarter-note delay in a send-and-return configuration to add sustain, rhythmic emphasis, and interest to a variety of track types including vocals and guitar solos.

- Get an Edge-like guitar sound (U2) by running a simple guitar part into a dotted-eighth-note delay synced to the song's tempo. To set this delay time in the DDL-1 device, use the 1/16 Step Length and set the delay time to 3. Set the Feedback knob high enough to give at least two to three repeats of each note.

Wet Versus Dry Signals

When working with time-based effects such as reverb and delay, you will need to balance the original signal with the processed signal. Audio producers refer to the original signal as the *dry* signal and the reverberated or delayed signal as the *wet* signal. Let's take a look at some common ways to adjust the wet/dry balance.

Using Time-Based Effects as Inserts

Although time-based effects are typically applied using a send-and-return configuration (see below), at times you may use a reverb or delay as an insert directly on a source Audio Track or Mix Channel device. For example, you may want to add reverb to a snare drum to give it a unique sound, or you may want to create a swirling guitar wash, where the signal is 100 percent wet. In these scenarios, you can place the desired device directly into the Insert FX container of the target Audio Track or Mix Channel device.

When applying a time-based effect directly to an Audio Track or Mix Channel devide, the only way to adjust the wet/dry balance is to use the device's controls. The location and labeling of these controls can vary from device to device. The RV7000 MkII and DDL-1 use knob controls labeled **Dry-Wet** and **Dry/Wet**, respectively.

Figure 8.23 RV7000 MkII's dry/wet balance is adjusted using a knob on the right side of the device.

The knob control can be rotated clockwise or counterclockwise to adjust the balance of wet and dry signals.

 As a general rule, you'll want to set the device to a relatively dry setting to emulate natural reverb or delay characteristics. Start with a low setting, such as 20 percent, and adjust from there as needed.

Using Send-and-Return Configurations

Configuring your time-based effects using a send and return is typically preferable, for a variety of reasons. First, you'll often want to apply the same reverb settings to many tracks in your project, so using a single device is much faster to set up and adjust.

Second, because time-based effects can use significant CPU power, using a send and return will save on CPU processing. You can create one instance of a reverb or delay on the main mixer's Master Section and then send audio to the effect from as many mixer channels as necessary to apply the effect to multiple sources.

And third, each source track's direct output will remain completely dry, while the send level can be used to adjust the wet signal. This makes it quick and easy to modify the wet/dry balance for a whole bunch of tracks in your project without needing to change the settings of the device itself.

Although advanced send-and-return configurations are beyond the scope of this book, you can configure a basic setup with a few simple steps. Use the techniques outlined below to get started with a reverb send.

 The same techniques can be used to create a delay send. Simply change the device you add to the rack.

To set up a reverb send effect:

1. Choose **CREATE > CREATE SEND FX**. Reason will display the Browser if it isn't already visible, set the browse focus to the **Create Send FX** context, and navigate to the Effects location.

2. Do one of the following:

   • Select the RV7000 MkII Reverb device and click the **CREATE** button at the bottom of the browser.

   • Double-click the RV7000 MkII Reverb device.

   A new RV7000 MkII device will be connected to a set of FX Send/FX Return jacks on the Master Section device in the upper-left portion of the Racks view.

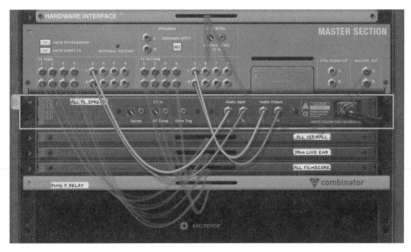

Figure 8.24 The reverb device connections to the FX Send/Return jacks on the Master Section

 The default empty song template in Reason Intro includes some send effect devices that are already connected to the Master Section, as seen in Figure 8.24.

Send effects in the Main Mixer view are identified by numbered positions 1 through 8. To use a send effect, you need to first determine which number corresponds to the effect. You can figure out the number of the effect you want to use by looking at the FX Send or FX Return area of the Master Section strip in the Main Mixer view.

The Master Section in the main mixer is always found on the right side of the view. The FX Send and FX Return areas have sets of numbered controls corresponding to send effects; alongside these controls Reason displays the name of the send effect device from the rack, so you can match up the effect device name to its send number. You can also determine the number by looking at the numbered set of jacks that the effect connects to in the Racks view, as seen in Figure 8.24 above.

If you start with the default empty song template, the new reverb effect will be effect number 5. The effect name will be **ALL Plate Spread**, matching the default patch loaded for the RV7000 MkII device.

Figure 8.25 The FX Return area of the Master Section strip in the Main Mixer view. Note the ALL Plate Spread effect in FX Return position 5.

To send audio to the reverb effect from a source channel, do the following:

1. In the Main Mixer view, locate the mixer channel strip for the audio track or Mix Channel to which you would like to apply the reverb effect.

2. Locate the Send section of the channel strip.

3. Click the button in the Send section corresponding to the send effect you want to use to activate that send position for the channel strip.

4. Adjust the **LEVEL** knob next to the numbered button to change the level of the audio going to the send effect and to blend the effect with the dry signal.

Figure 8.26 Sending audio from a mixer channel to the fifth send effect.

 The Master Section in Reason Intro supports up to eight send effect connections. After eight connections are used, the Create > Create Send FX menu item will be disabled. You can remove a send effect and free up connections by deleting a send effect device in the Racks view.

Review/Discussion Questions

1. How can you access the Insert FX container for an Audio Track or Mix Channel device in the Racks view? (See "Device Basics" beginning on page 202.)

2. How can you access the Insert FX container for an Audio Track or Mix Channel device from the Main Mixer view? (See "Device Basics" beginning on page 202.)

3. How can you insert an effect device on an Audio Track or Mix Channel device? How can you remove an inserted effect device? (See "Inserting Effects on Audio Track and Mix Channel Devices" beginning on page 204.)

4. How can you move an effect device from one Audio Track or Mix Channel device to another? How can you duplicate an effect device? (See "Moving and Duplicating Effects Devices" beginning on page 208.)

5. How do you open the VST plug-in window for a plug-in rack device? (See "Displaying VST Plug-In Windows" beginning on page 209.)

6. How does the Power/Bypass switch work on effect devices? (See "Common Device Controls" beginning on page 215.)

7. What is the purpose of the Select Patch buttons in a device's Patch Section? (See "Common Device Controls" beginning on page 215.)

8. Which keyboard modifier can you use to make fine adjustments on device controls? (See "Device-Specific Parameters" beginning on page 217.)

9. Which keyboard modifier can you use to reset a device control to its default value? (See "Device-Specific Parameters" beginning on page 217.)

10. What are the four common types of EQ bands? (See "Types of EQ" beginning on page 218.)

11. What two EQ effect options are included with Reason Intro? (See "EQ Effects in Reason Intro" beginning on page 219.)

12. What four types of dynamics effects are found in Reason Intro? (See "Dynamics Effects in Reason Intro" beginning on page 224.)

13. What is the difference between reverb and delay? What category of effects do both of these processes belong to? (See "Reverb and Delay Effects" beginning on page 226.)

14. What are some scenarios in which you might place a time-based effect directly on an Audio Track or Mix Channel device? (See "Using Time-Based Effects as Inserts" beginning on page 229.)

15. What are some advantages to using a send-and-return configuration for time-based effects? (See "Using Send-and-Return Configurations" beginning on page 230.)

 To review additional material from this chapter and prepare for certification, see the Reason Audio Production Basics Study Guide module available through the Elements|ED online learning platform at ElementsED.com.

Optimizing Tracks with Signal Processing

🎧 Activity

In this exercise, you will learn how to use effects to optimize tracks in Reason. You'll begin by looking at some practical applications for equalization (EQ). Then, you'll explore some options for controlling dynamics. Finally, you'll create a send-and-return configuration to apply reverb for multiple tracks.

🕐 Duration

This exercise should take approximately 20 minutes to complete.

✛ Goals/Targets

- Use the Channel Strip EQ to cut bass frequencies and boost high frequencies
- Use the MClass Compressor device to reduce variations in peak volume levels
- Set up a reverb device in a send-and-return configuration
- Apply reverb processing using the RV7000 MkII device

Exercise Media

This exercise uses media files taken from the song, "Lights," provided courtesy of Bay Area band Fotograf.

Written by: Zack Vieira and Eric Kuehnl; Performed by: Fotograf

The media provided for this course may be used for educational purposes only. No rights are granted to use the media for any other personal, commercial, or non-commercial purposes.

Getting Started

To get started, you will open your completed project from Exercise 7. This will serve as the starting point for this exercise.

Open your existing Lights project:

1. Select **FILE > OPEN** or press **COMMAND+O** (Mac) or **CTRL+O** (Windows). The Browser will open.

2. Use the Browser to navigate to the folder containing your **Lights-xxx** project.

3. Select your **Lights-xxx** project from the list of available files.

4. Click **OPEN** at the bottom of the Browser. The project will open as it was when last saved. If needed, reopen the Main Mixer view (**WINDOW > VIEW MAIN MIXER**).

Applying EQ and Dynamics to Tracks

To get an idea of where EQ processing might be useful, you should take a moment to listen to the project in its current state. Pay particular attention to any areas where frequencies are clashing between multiple tracks. Also listen for tracks that could benefit from more controlled dynamics.

Evaluate the tracks:

1. Press the **SPACEBAR** to begin playback.

2. Listen for tracks that compete for space in the same frequency range, such as the **01 Kick** track and the **05 Synth Bass** track. Try to formulate some ideas about how to give each track its own space in the frequency spectrum.

3. Also, listen for tracks with dynamics that might cause problems. Listen for consistency in the volume of notes for tracks that should serve as a steady backbone, such as the bass guitar during the choruses.

4. Feel free to solo and mute tracks to isolate certain elements. And move the faders to see if you can find a good balance of tracks, or if certain tracks present unique problems.

   • Is it hard to separate certain tracks from one another without causing one to obscure the other in a given frequency range?

   • Is it hard to find a level for certain tracks that allows all of the notes to be heard without causing certain notes to get too loud?

Adding EQ Processing

When using EQ, you may find yourself wanting to boost frequencies to emphasize the best parts of a track. However, it is equally important to cut frequencies to create room for other tracks to be heard. In this section of the exercise, you will use EQ to reduce the bass frequencies in the 05 Synth Bass track while slightly boosting the high frequencies, to separate the Synth Bass from competing bass frequencies in the 01 Kick track.

Navigate to the Channel Strip EQ controls for the Synth Bass track:

1. Locate the channel strip for the 05 Synth Bass track in the mixer.

2. If necessary, scroll the Main Mixer view vertically to find the EQ section for the Synth Bass channel.

Adjust the EQ settings:

1. Turn on the EQ bands that you expect to use. The default settings have all bands disabled.

    *   Enable the HPF band by clicking the **ON** button for the band so that it becomes lit in blue.

    *   Enable the primary group of four EQ bands by clicking the **ON** button just above the **E** button so that it becomes lit in blue.

Figure 8.27 Clicking the On button for the group of four EQ bands (shown inactive)

2. Set the parameters for each EQ band:

    *   **HPF:** Set the frequency (Hz knob) around 65.0 Hz.

    *   **LF:** Set the frequency (Hz knob) to around 200.0 Hz and the gain (dB knob) to around –3.0 dB.

    *   **LMF:** Set the Q to around 1.25, the frequency (kHz knob) to around 800.0 Hz, and the gain (dB knob) to around –2.0 dB.

    *   **HF:** Set the frequency (kHz knob) to around 5.00 kHz and the gain (dB knob) to around 2.0 dB.

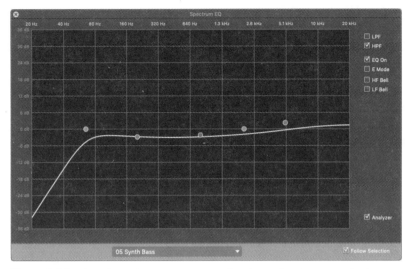

Figure 8.28 The Channel Strip EQ settings for the 05 Synth Bass track, shown in the Spectrum EQ window

After configuring the EQ, you may notice that the 05 Synth Bass track has become a bit quieter in the mix, which occurs because of the frequency cuts made by the EQ. To compensate for the reduction in volume caused by the EQ, try bringing the Synth Bass channel fader up by about 2.0 dB from its current position.

Adding Dynamics Processing

Next you'll apply dynamics processing to help control the signal levels. The 08 Bass Gtr track has a slightly wider dynamic range than ideal, making it harder to position in the mix without constantly adjusting the Volume fader. Here you will rein in the dynamic variation using a compressor.

Insert a compressor on the Bass Gtr track:

1. Click on the 08 Bass Gtr channel in the mixer to select it.

2. Open the Browser and navigate to the Effects location.

3. Click on the MClass Compressor device in the Browser List to select it.

4. Click the **CREATE** button at the bottom of the Browser to add the compressor to the Insert FX container of the 08 Bass Gtr track. Reason will display the compressor in the Racks view.

Configure the compression settings:

1. Set the Threshold to around −35.0 dB.

2. Set the Ratio to around 2.5:1.

3. Set the Attack to around 7ms.

4.   Set the Release to the maximum value (600ms).

5.   Set the Output Gain to around 10.0 dB.

Figure 8.29 The MClass Compressor settings for the 08 Bass Gtr track

Listen to the results:

1.   Use the Sequencer view to go to the first chorus of the song at Bar 49.

2.   Play through the first chorus of the song and listen to the **08 Bass Gtr** track. The dynamic variation should be dramatically reduced.

3.   Continue to adjust the Output Gain control on the MClass Compressor to find a good volume level for the track.

Using a Send-And-Return Configuration for Reverb

In this section, you'll add some reverb to help smooth out the strings and make them sound more natural. You'll set up a send-and-return configuration that can be used by multiple tracks in the project.

Add a new reverb send effect and load a reverb patch file:

1.   Click **CREATE > CREATE SEND FX**.

2.   In the Browser, select the **RV7000 MkII REVERB** device.

3.   Click the **CREATE** button at the bottom of the Browser to add the reverb device as a Master Section send effect.

     The Browser will set a browse context for the new reverb device and will display a list of patch files. (If you don't see the list of patches, you can navigate to the correct location by going to **FACTORY SOUNDS > RV7000 MkII PATCHES** in the Browser.)

4.   Double-click the **ALL HEALINGSHALL.RV7** patch file to load it.

Figure 8.30 The RV7000 MkII device with the suggested settings

Send audio from the Strings track to the Reverb track:

1. Display the Main Mixer view if it is not already visible.

2. Identify the send effect number for the new reverb effect by looking for the **ALL HealingsHall** device name in the FX Return area of the Master Section strip.

Figure 8.31 The FX Return area in the Master Section strip

3. Locate the Send section on the **Strings** channel.

4. Click on the numbered send effect button identified in Step 2.

5. Increase the level of the send to about **–5 dB**. You should now be able to hear the reverberated strings when you play the second chorus of the project at Bar 77.

Figure 8.32 The send controls for the reverb send on the Strings channel

Finishing Up

To complete this exercise, you will need to save your work and close the project. You will be reusing this project in Exercise 9, so it's important to save the work you've done.

Before you wrap up, you can also listen to the project to hear the processing changes you've made.

Finish your work:

1.   Choose **FILE > SAVE** to save the project.

2.   Press **[.] (PERIOD)** on the numeric keypad to move the Song Position Pointer to the beginning of the project.

3.   Press the **SPACEBAR** to begin playback and listen through the project. Press the **SPACEBAR** a second time when finished to stop playback.

4.   Press **[.] (PERIOD)** on the numeric keypad again to return the Song Position Pointer to the start of the project.

5.   Choose **FILE > CLOSE** to close the song.

That completes this exercise.

Chapter 9

Finishing a Project

...*What You Need to Do to Create a Stereo Mixdown...*

This chapter introduces you to the processes of adding automation and creating a stereo mixdown of your Reason song. We cover the process of recording automation in real time on your tracks. We also cover a variety of editing techniques that are available for working with the automation on your tracks. We then review uses for limiters in the Master Section. The chapter wraps up with a discussion on exporting your mix as a stereo file.

✪ Learning Targets for This Chapter

- Understand the settings that will and will not be recalled with a song

- Learn how to enable tracks for automation recording

- Learn how to write automation while recording

- Understand how to edit automation clips

- Add appropriate processing to the Master Section

- Use the Export Song as Audio File command to create a stereo mix of a project

Key topics from this chapter are illustrated in the Reason Audio Production Basics Study Guide module available through the Elements|ED online learning platform. Sign up at ElementsED.com.

After you have finished balancing your tracks, adding processing to the tracks, and adding any desired effects for the mix, it's time to start thinking about finishing up the project and creating a stereo mixdown. As you put the finishing touches on your mix, you can use automation to create and save dynamic changes. Then you can export your mix as a stereo file, configuring the file parameters as needed.

Recalling a Saved Mix

When you open a song that you've previously worked on, all of Reason's mixer settings are recalled with the file, including the following:

- **Track layouts and views**—The track order, track heights, and zoom settings that were in place when the project was last saved will all be recalled.

- **Mix attributes, settings, and automation**—The settings for Volume faders and pan controls will be recalled for each mixer channel, along with any automation you previously created.

- **Track and send routing**—Any routing you've used for output busses, effects sends, and similar configurations will be recalled, along with the send levels and FX Return section settings.

- **Device assignments**—Insert effect device assignments and parameter settings will be recalled for each audio track and Mix Channel, as well as the settings for all other devices in the rack.

Occasionally, however, conditions may exist that will prevent a song from recalling exactly as it was when last saved. These conditions can include the following:

- **External effects processors**—As discussed in Chapter 7, if you use external gear for your mix, you will need to reconnect the gear and reconfigure all the hardware settings whenever you return to the project.

- **Unavailable plug-ins**—If you open the project from a system that does not have the same plug-ins installed as the system where you created the project, any unavailable plug-ins in the project will not be accessible. Missing Rack Extension plug-ins will be replaced by "Missing Device" placeholders in the rack, while VST plug-in rack devices for missing VST plug-ins will be empty and will display the words "Missing Plugin" when unfolded.

- **Unavailable samples and loops**—All of the audio and MIDI that you record in the sequencer is saved with your song file, so you don't have to worry about that content being inaccessible. However, audio samples and REX loop files loaded by instrument devices such as the Kong Drum Designer are not stored with your song file. Instead these are loaded from their original sound library or folder locations on your computer. If the audio samples are not available on a computer when you open the song, Reason will display a prompt for you to find the missing sounds.

Reason provides a feature called Song Self-Contain Settings that allows you to optionally specify audio samples and loops to save into your song file so that the audio data remains available when you open the song on another system.

Using Automation

On a simple project, you may be able to set the Volume faders, pan controls, and other parameters on mixer channels and devices and leave them unchanged from the start of the mix to the end. But often a mix will require dynamic changes during the course of playback. This is where automation can be useful.

Reason allows you to record changes you make to parameters as automation on mixer channels and rack devices. Automatable parameters include mixer volume, pan, and mute controls; send controls; and most device controls.

Automation, Tracks, and Lanes

Reason stores parameter automation information on automation lanes that are associated with tracks in the Sequencer view. For example, an audio track in the sequencer will have a single lane for audio data, but it can also have multiple automation lanes that are used to change parameters dynamically throughout a song. Similarly, an instrument track will have one or more note lanes for note information, along with optional automation lanes. For other devices, such as Mix Channel devices and effects devices, you can create dedicated automation tracks to automate the device parameters.

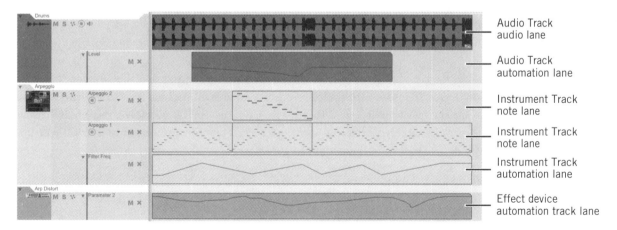

Figure 9.1 Tracks and lanes in the Sequencer view

One aspect of automation that can be a bit confusing at first is the track setup required to automate parameters in the main mixer, such as the Volume fader level and pan controls. For audio tracks, the process is straightforward. As discussed in earlier chapters, creating an audio track in the sequencer also creates an Audio Track device in the rack, and the Audio Track device is linked to a corresponding mixer channel in the Main Mixer view. As a result, for audio tracks, you don't need an additional track to automate the mixer settings; mixer automation is recorded to automation lanes associated with the audio track. (See Figure 9.1.)

For Instrument devices, the situation is more complicated, though. It might seem that mixer settings such as volume and pan controls can be automated using the instrument track, but unfortunately that is not the case. Recall that instrument devices in the rack are cabled to Mix Channel devices. It is the Mix Channel device that is associated with the mixer channel in the Main Mixer view, not the instrument device. To automate the mixer channels linked to your instruments, you actually need to automate the Mix Channel devices.

Figure 9.2 Main mixer channel, the corresponding Mix Channel device, and the connected Instrument device

To automate a Mix Channel device, the device needs to have an automation track in the sequencer. When you create an instrument device, Reason automatically creates an instrument track, then creates a Mix Channel device and connects the instrument device to the Mix Channel device. However, the Mix Channel device itself does not automatically receive a track in the sequencer. Therefore, you need to manually create a sequencer track for any Mix Channel device you want to automate.

To create a Mix Channel device automation track:

1.   Click on the Mix Channel device in the Racks view to select it.

2.   Do one of the following:

   • Choose **EDIT > CREATE TRACK FOR <DEVICE NAME>**, where **<DEVICE NAME>** is the name of the Mix Channel device.

- Right-click on the background of the Mix Channel device outside of any of the controls or parameters. Choose **CREATE TRACK FOR <DEVICE NAME>** from the pop-up menu, where **<DEVICE NAME>** is the name of the Mix Channel device.

Reason will create an empty automation track for the Mix Channel device.

You can also create an automation track for the Mix Channel device by doing one of the following:

- Right-click on a parameter or control in the Mix Channel device or the mixer channel strip. Choose **EDIT AUTOMATION** from the pop-up menu.

- **OPTION-CLICK** (Mac) or **ALT-CLICK** (Windows) on a parameter or control in the Mix Channel device or the mixer channel strip.

Reason will create an automation track for the Mix Channel device and will simultaneously create an automation lane for the selected parameter.

Figure 9.3 A Mix Channel automation track with an automation lane for the main Volume fader (Level)

 You can also use the techniques described here to create automation tracks for other types of devices, such as effects devices and utility devices.

Recording Real-Time Automation

Once you have identified or created the tracks you want to automate, you can begin the process of recording real-time automation. Recording real-time automation means that you can make changes to parameters while your song is recording, and Reason will capture the changes you make as automation data. The recorded parameter changes will then play back as part of the song.

To record parameter automation for a track, the track must be record-enabled for parameter automation. Record-enabling a track for parameter automation is similar to record-enabling a track for audio or MIDI recording. The track header for each track in the Track List contains a Record Enable Parameter Automation button that controls whether the track is enabled for automation recording.

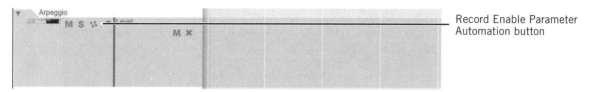

Record Enable Parameter
Automation button

Figure 9.4 An automation track and the Record Enable Parameter Automation button

To enable or disable a track for parameter automation, click the RECORD ENABLE PARAMETER AUTOMATION button to toggle it on or off. You can have any number of tracks simultaneously enabled for parameter automation.

As described in Chapter 6, the default settings in Reason record-enable a track whenever you select it in the Sequencer view. This behavior extends to record-enabling parameter automation as well: simply selecting a track will enable parameter automation recording. However, the parameter automation record-enable state can still be controlled independently, so you can also use the Record Enable Parameter Automation button to turn it on or off.

 If you want to use a MIDI controller to adjust parameters while recording automation, in most cases you will still need to set Master Keyboard Input to the target track.

Similarly, the MANUAL REC button in the Sequencer view (which prevents Reason from automatically record-enabling selected tracks) also applies to the parameter automation record-enable mode for tracks. When the Manual Rec button is activated, Reason will not automatically record-enable parameter automation when you select a track.

Avoiding Accidental Recording

Reason's default settings allow you to start recording very easily just by selecting a track in the sequencer. However, after you've finished recording and are preparing to work with automation, it is possible to accidentally record MIDI or audio when you intended to record only automation.

For example, imagine you've recorded a number of audio tracks, and now you want to record automation. If you click on an audio track in the sequencer to select it (or move Master Keyboard Input to the audio track in the rack), the audio track will be enabled both for automation recording and for audio recording. If you start a record pass to record automation, you'll also start recording audio onto the audio lane, which is probably not what you intended.

So when preparing to record automation for an audio track or instrument track, it is important to turn off the standard Record Enable button if you don't intend to record media along with the automation. The Manual Rec feature in the sequencer can be very helpful in these cases because it requires you to manually record-enable the type of input you want to record.

To automate basic settings, do the following:

1.   Put the target track(s) into Record Enable Parameter Automation mode.

2.   Begin a record take. Automation will not be writing at this point.

3.   When you reach the location where you want to add automation, adjust the volume, pan, or other settings on the associated device(s) or mixer channel(s). The changes will be recorded as automation data. Reason automatically creates automation lanes for each parameter you change.

4.   Continue the record take through the song, adjusting the settings as needed.

5.   When you are finished adding automation, stop the record take.

The Automation Override Indicator

The Transport Panel contains an important indicator called the Automation Override indicator.

Figure 9.5 The Automation Override indicator in the Transport Panel.

Automation Override Conditions

The Automation Override indicator will light under two main conditions. First, the indicator will light after you finish an automation record pass where you have modified some parameters. In this situation, the indicator signifies that the parameters you changed are holding their positions at the last values you set them to, rather than reflecting one of the recorded automation values.

Second, when working with a song that has existing automation, you can "take over" one of the automated parameters using your mouse or MIDI controller. The Automation Override indicator will light to show that the parameter has temporarily stopped following automation and will remain at the value you set.

Put simply, the Automation Override indicator lights up when at least one automated parameter has been modified and is not following its recorded automation.

Clearing Automation Overrides

You can clear overrides and reset the Automation Override indicator by doing one of the following:

■   Click on the **AUTOMATION OVERRIDE** indicator itself to clear it.

■   If the transport is currently playing back, stop the transport and restart playback to clear the Automation Override indicator.

■   If the transport is stopped, click the **STOP** button again to clear the Automation Override indicator.

The Automation Override indicator is also related to another aspect of visual feedback that Reason provides to help you manage automation in your project. Every automated parameter in the rack or the main mixer is surrounded by a green border when the parameter is following its recorded automation. When automation is overridden for an automated parameter, the green border will not be displayed.

Figure 9.6 A Europa instrument device with three automated parameters indicated by green borders

Working with Automation Clips

Each automated parameter on a device or mixer channel has an associated automation lane in the Sequencer view. Automation lanes store automation data in clips, allowing you to see a visual representation of the automation changes. You can perform standard editing techniques on automation clips, including the following:

■   Moving

■   Resizing

■   Deleting

■   Cutting/copying/pasting

■   Creating new clips with the Pencil Tool

■   Erasing with the Eraser Tool

- Splitting with the Razor Tool

- Muting with the Mute Tool

These editing functions are applied to automation clips in the same way that they would be applied to audio or note clips. You can also open automation clips for editing, which allows you to change the automation points that define the shape of the automation changes.

 See Chapter 6 for information on clip editing techniques.

Figure 9.7 An effect device automation track with a lane containing an automation clip

Editing Automation Points

With an automation clip available on a parameter automation lane, you can edit the automation using point-editing techniques. Automation points (also called automation events) are points in an automation clip where the automation line changes slope, shape, or direction.

Opening an Automation Clip

Before you can edit automation points, you first need to open an automation clip.

To open an automation clip for editing, do one of the following:

- Double-click the automation clip with the Selection Tool.

- Click on the automation clip to select it and then press **RETURN** or **ENTER**.

- Click on the automation clip to select it. Then, click the **EDIT INLINE** edit mode button in the Sequencer view toolbar.

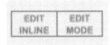

Figure 9.8 Automation clip edit mode buttons in the Sequencer view toolbar

When you open an automation clip in this way, Reason provides an inline editing experience that lets you view and edit automation points directly in the Arrangement pane of the Sequencer view. The automation lane containing the open clip will expand in height, and the individual automation points will become visible. Some additional controls also appear in the Track List area for the automation lane.

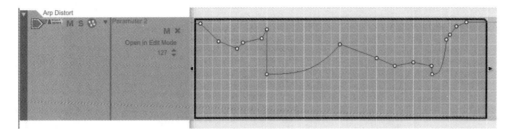

Figure 9.9 An automation clip open for inline editing in the Sequencer view

 Reason offers another edit mode button for automation clips, labeled Edit Mode. Using this button swaps the Arrangement pane with a full-screen editor for the track. This book focuses on using inline editing for automation clips.

To close an automation clip after opening it for editing, do one of the following:

■   Press **ESCAPE**.

■   Click on any other lane in the sequencer.

Using the Selection Tool for Automation

You can edit the automation clip by adding, moving, or deleting points using the Selection Tool.

To edit automation with the Selection Tool, do any of the following:

■   Double-click in the automation clip to add a single point.

■   Double-click and drag in the automation clip to add multiple points.

■   Click an existing point to select it.

■   Click an empty area of the clip and drag to draw a selection rectangle and select multiple points.

■   Click and drag an existing point horizontally to adjust its position or vertically to adjust its value.

■   Double-click on an existing point to remove it.

■   Press **DELETE** or **BACKSPACE** to remove selected points.

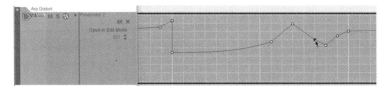

Figure 9.10 Editing automation points using the Selection Tool

The Selection Tool works with the Snap function. If the Snap function is enabled when you add automation points, the new points will snap to the nearest grid line. Likewise, if the Snap function is enabled when you move automation points horizontally, they will move in steps according to the current grid size value.

By default, automation clips consist of straight lines between points. You can use the Selection Tool to create curved shapes between two points. You can also reset a curved segment to set it back to a straight line.

To create curved automation shapes, do the following:

1.   Position the Selection Tool over the line between two automation points (not over the automation points themselves). The mouse cursor will change to show a curve tool.

2.   Click and drag up or down to change the shape of the automation line.

You can change the shape of an existing curved line by repeating the steps above.

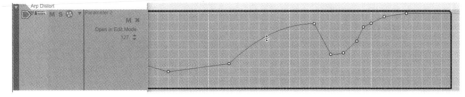

Figure 9.11 Creating a curved automation shape

To reset a curved shape back to a straight line:

1.   Position the Selection Tool over the curve.

2.   Hold **SHIFT** and click on the curve.

Automation playlists can also be edited using other sequencer tools, such as the Pencil Tool.

Using the Pencil Tool for Automation

The Pencil Tool can be useful for drawing automation changes in an automation clip.

To edit automation with the Pencil Tool, do any of the following:

- Click and drag to draw multiple points, similar to double-clicking and dragging with the Selection Tool.

- Click and drag while holding **OPTION** (Mac) or **CTRL** (Windows) to draw a horizontal automation line that has the same value at the start and end points.

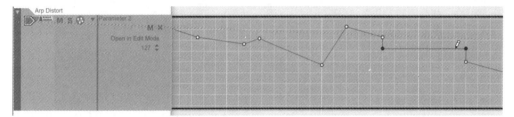

Figure 9.12 Drawing a horizontal automation region with the Pencil Tool

The Pencil Tool also works with the Snap function. If the Snap function is enabled, points drawn with the Pencil Tool will snap onto the grid lines. No points will be added between grid lines as you draw points with the Snap function enabled.

Using the Eraser Tool for Automation

The Eraser Tool can also be used on an automation clip. To remove points with the Eraser Tool, click and drag in the automation clip to draw a selection rectangle around one or more automation points. When you release the mouse button, all points contained in the rectangle will be deleted.

MIDI Performance Controllers

When you record MIDI performances on instrument tracks, some controls that you can use while playing the instrument are identified by Reason as belonging to the set of standard MIDI performance controllers. These standard controls include Pitch Bend, Mod Wheel, Sustain, Aftertouch, Breath Control, and Expression. Of these, Pitch Bend and Mod Wheel controllers are the most commonly used.

Performance controllers can cause a bit of confusion because they work alongside parameter automation, with a few differences:

- Performance controllers apply only to instrument tracks.

- Performance controllers are recorded into note clips on note lanes, not onto automation lanes.

- Performance controllers are not edited inline. They are edited by opening note clips in edit mode.

- Performance controllers can be recorded on instrument tracks that are not enabled for parameter automation recording. The instrument track only needs to be enabled for note recording.

For example, if you use the Pitch Bend control on your keyboard while recording a performance, the Pitch Bend control on the instrument device in the rack will have a green border, indicating that the pitch bend is automated.

Figure 9.13 A Europa instrument device with a green automation border around the Pitch control at the bottom left

However, when you check the sequencer, the instrument track will not have an automation lane for the Pitch Bend parameter. Instead, the Pitch Bend controller automation is recorded into a note clip.

Figure 9.14 A note clip with performance controller automation

Parameter Automation Static Values

Because Reason stores parameter automation in discrete clips on automation lanes, there will often be gaps between clips where the automation lane doesn't contain any automation data. When playback reaches these points, automated parameters fall back to a default value called the *static value*.

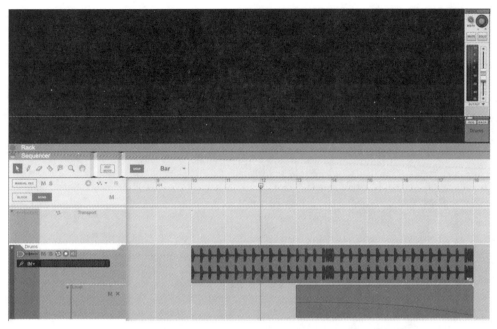

Figure 9.15 An audio track with automation applied to the level fader

In Figure 9.15, automation has been applied to the fader for an audio track starting at Bar 13 so that the audio track fades out by the end of the audio clip at Bar 18. Outside of Bars 13 to 18, though, there is no automation on the Level automation lane. When the Song Position Pointer is outside of any automation clip, as seen at Bar 12, the fader for the audio track sits at a default position defined by the static value.

When you first add automation to a parameter such as a fader, Reason captures the initial value of the parameter. That initial value becomes the static value for the parameter. A good general practice is to set your parameter controls to the basic positions you want before you start adding automation. When Reason plays back sections of your song that don't contain any automation clips, the parameter controls will revert to the default positions that you set before adding automation.

You can also change the static value for any parameter after automation has been added:

1. Open any automation clip for inline editing on the automation lane for the target parameter.

2. Locate the static value controls in the Track List header for the automation lane, just below the button labeled **OPEN IN EDIT MODE**.

3. Do one of the following to change the static value:

   • Click on the static value itself and drag it up or down.

   • Click on the up/down arrows next to the static value to change the value.

   • Double-click on the static value to open a text box where you can type in a new value.

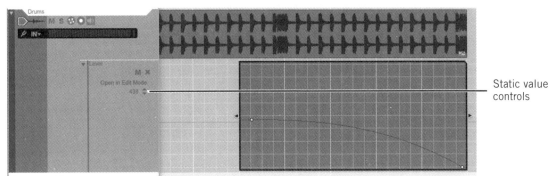

Figure 9.16 Static value controls displayed in the Track List for an automation lane

 The format and range of values that can be selected for a static value will vary for different types of parameters.

Muting and Clearing Automation

At times you may want to prevent Reason from playing back automation clips, or you may want to remove automation for a parameter entirely. One way to do so would be to apply standard clip editing techniques. You can mute and delete automation clips in exactly the same way as audio or note clips. However, Reason also provides some other tools to manage parameter automation.

Every automation lane has both Mute and Delete buttons in the Track List area for the lane:

- Click the **MUTE** button to prevent playback of all automation data on the automation lane. The parameter associated with the lane will remain at the value currently in effect at the time you mute the lane.

- Click the **DELETE AUTOMATION LANE** button to delete the lane entirely and remove automation for the parameter. If the automation lane contains at least one clip, Reason will prompt you to confirm deletion of the lane.

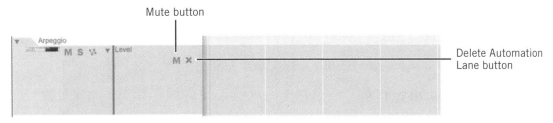

Figure 9.17 Parameter automation lane buttons

You can also remove all automation for a parameter by doing the following:

1. Locate the automated parameter control in the Racks view or Main Mixer view.

2. Right-click the control and choose **CLEAR AUTOMATION** from the pop-up menu.

Creating a Mixdown

Mixing down is the process of recording the output from multiple tracks to a stereo file. This process is also commonly referred to as *bouncing* the project. Mixing down is often the last phase of audio production, although you can create a bounce at any time to create a complete mix as a stereo file.

Adding Processing on the Master Section

Prior to creating a final mix, you may want to consider using some final dynamics processing on the entire mix to optimize your output levels. You can also use dither during a bounce to preserve the low-level dynamics in your mix during a bit-depth reduction.

Adding a Limiter to Optimize Output Levels

As discussed in Chapter 8, a limiter is a dynamics processor that caps the audio signal at a specified level. A limiter can be used in the Master Section to protect against clipping. Using a limiter can also help make the overall mix seem louder. Here, we will use the MClass Maximizer processor as a brickwall limiter.

If you create a song from one of the templates provided with Reason, you will have a mastering chain of effects devices already loaded into the Insert FX container of the Master Section device. For this discussion, we assume the Master Section is empty; we will add only the MClass Maximizer insert device.

 The Reason Preferences settings include an option to select a template to use when creating a new song project. The default setting uses the Empty + FX template.

If needed, you can clear the initial insert effects from the Master Section, as follows:

1. Locate the Master Section device in the upper-left area of the Racks view.

2. Click on the Master Section device to select it.

3. Do one of the following to remove the existing insert effects devices:

    • Choose **EDIT > CLEAR INSERT FX**.

    • Right-click on the background area of the Master Section device and choose **CLEAR INSERT FX** from the pop-up menu.

The Insert FX container of the Master Section is bypassed in most templates. In this state, any insert effects you add will not perform any processing. To hear the effect of the MClass Maximizer, you will need to make sure that the Bypass Master Insert FX button is not active. You can do this from the Racks view or from the Main Mixer view.

To enable the Master Section insert effects from the Racks view, do the following:

1. If the Master Section device in the Racks view is folded, click the **FOLD/UNFOLD** button (triangle) to unfold it.

2. If needed, click the **BYPASS MASTER INSERT FX** button on the Master Section device to deactivate it so that it is no longer lit with a blue background.

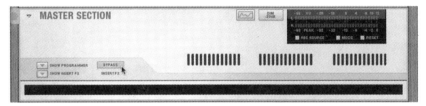

Figure 9.18 The Bypass Master Insert FX button (shown active)

To enable the Master Section insert effects from the Main Mixer view, do the following:

1. Locate the Master Section strip on the right side of the view.

2. Go to the Master Inserts area of the Master Section strip.

3. If needed, click the **BYPASS MASTER INSERT FX** button at the top of the Master Inserts area to deactivate it so that it is no longer lit with a blue background.

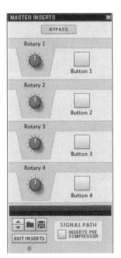

Figure 9.19 The Bypass Master Insert FX button at the top of the Master Inserts area in the Main Mixer view

With the Master Section empty and the Bypass Master Insert FX button switched off, it is time to add a limiter to your mix and configure it to optimize your overall mix levels.

To add a limiter to your mix in Reason, do the following:

1. Unfold the Master Section device in the Racks view, if needed.

2. Click the **SHOW INSERT FX** button to open the Insert FX container for the Master Section.

3. Access the Browser and navigate to the Effects location.

4. Drag and drop the **MCLASS MAXIMIZER** device to the Master Section Insert FX container.

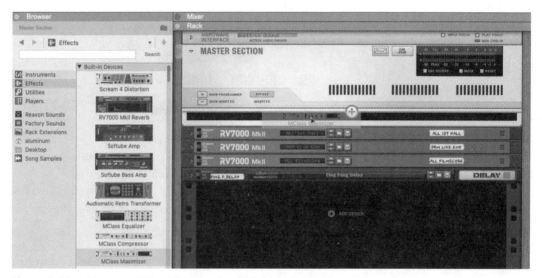

Figure 9.20 Adding an MClass Maximizer to the Master Section Insert FX container

To configure the limiter to optimize your mix levels, do the following:

1. Activate the button labeled **4MS LOOK AHEAD** on the MClass Maximizer device. This is required for the device to function as a brickwall limiter.

2. Leave the **ATTACK** and **RELEASE** buttons at their default **Fast** settings to provide a quick response for the momentary peaks you are targeting with this type of limiter. The Fast attack time is also required for the MClass Maximizer device to function as a brickwall limiter.

3. Play through the project, taking note of the gain reduction readings displayed on the Gain meter. Assuming that your mix was not clipping prior to adding the limiter, you should see little activity on the meter.

4. Turn up the **INPUT GAIN** control to increase the level of the signal flowing into the limiter. Doing so will increase the overall loudness of the mix while still keeping the peak levels from clipping. Try

starting with a value of around 3.0 dB. Your goal should be to increase the loudness a bit while not really hearing any other changes to the sound from the limiter. If you increase the gain too much, you will start to notice undesirable compression artifacts, such as "pumping" or "breathing" effects.

Figure 9.21 The MClass Maximizer with the Input Gain and 4ms Look Ahead controls set

Creating a CD-Ready Mix

To create a CD-ready mixdown from Reason, you will need to perform a bit-depth reduction: your mixdown file will need to use 16-bit audio in order to be burned onto a CD.

 The Red Book audio CD standard requires audio files that are encoded with a sample rate of 44.1 kHz and a word length of 16 bits.

As discussed in Chapter 3, lower bit-depth audio also exhibits a reduced dynamic range. To help preserve the dynamic range when reducing bit depth, you can add *dither* while bouncing an audio file. Dither is a form of randomized noise used to minimize signal loss when audio is near the low end of its dynamic range, such as during a quiet passage or fade-out.

Adding Dither to a Mix

Proper use of dithering allows you to squeeze better subjective performance out of 16-bit audio.

In Reason, dither is applied during the process of exporting audio. During this process, you will be prompted to choose the output bit depth of the audio file and allowed to select whether to apply dither. To create a mix for CD, you would enable the Dither checkbox to optimize the final 16-bit output.

Figure 9.22 The Audio Export Settings dialog box

If you will be keeping your mixed file at 24-bit, you should *not* use dither. Reason will disable the Dither checkbox when a Bit Depth value of 24 is selected.

 Third-party plug-in devices can also be used to add dither. If you use such a device in the Master Section with dither enabled, make sure that you do not use the Dither checkbox in the Audio Export Settings dialog box. Otherwise you will be applying dither twice and adding unnecessary noise.

Considerations for Bouncing Audio

When exporting a mix from Reason, the bounced file will capture all audible information in your mix just as you hear it during playback.

The following principles apply to the mixdown file:

- **The file will include only audible tracks.** What you hear during playback is exactly what will be included in the mix. Any tracks that are muted will not be included. Similarly, if any tracks are soloed, they will be the *only* tracks included.

- **The file will be a rendered version of your project.** Inserts, sends, and external effects will be applied permanently. Prior to exporting the mix, listen closely to your entire project to ensure that everything sounds as it should. Pay close attention to levels, being sure to avoid clipping.

- **The file will be created based on the Song End Marker or Loop Locators.** If you use the FILE > EXPORT SONG AS AUDIO FILE command when you export your mix, the mix file will extend to the position of the Song End Marker in the Ruler. If you use the FILE > EXPORT LOOP AS AUDIO FILE command when you export your mix, Reason will create a bounce from the area between the left and right Loop Locators in the Ruler.

Setting the Song End Marker

Before exporting your song as an audio file, you will need to ensure that the Song End Marker is set to an appropriate position. The Song End Marker defines the location where the audio export will stop.

 The Song End Marker is sometimes referred to as the End Position Marker or simply the End Marker.

The Song End Marker is displayed in the Sequencer view's Ruler and looks like a rectangle containing the letter E. (See Figure 9.23.)

To move the marker, simply click on it in the Ruler and drag it. If Snap is activated, the Song End Marker will follow Snap behavior and will jump between grid increments as you drag.

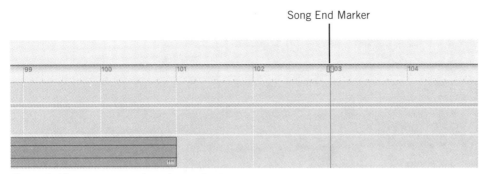

Song End Marker

Figure 9.23 The Song End Marker in the Sequencer view

Reason doesn't display grid lines in the Arrangement pane to the right of the marker, which provides a visual indication of where the song will stop during export. It is possible to position the Song End Marker so that there are clips beyond the end point. You can still play these clips in the sequencer, as the End Marker doesn't affect playback within the sequencer. However, nothing beyond the Song End Marker will be included when you export the song.

When positioning the Song End Marker, you will generally want to avoid putting the marker directly against the edge of the last clip in the sequencer. If your song includes any time-based effects, such as reverbs and delays, you will need to allow time for their signals to fade out after the last clip ends. If any effects are still audible at the point of the Song End Marker, they will be cut off in the mixed audio file.

 Try placing the Song End Marker a bar or so after the last clip ends. Be sure to listen to the project and ensure that any reverb and delay tails have completely faded to silence when playback reaches the Song End Marker.

Exporting an Audio Mix

When you are ready to create your mixdown in Reason, you can use the FILE > EXPORT SONG AS AUDIO FILE command. This provides a fast way to create a bounce to a stereo file, requiring little to no setup.

Selecting Export Options

The Export Song as Audio File command combines the outputs of all currently audible tracks to create a new audio file on your hard drive. You can select options for the exported file using the resulting dialog boxes.

To bounce all currently audible tracks, do the following:

1.   Verify that the project plays back as desired—check levels, processing, and mute/solo states for tracks— and ensure that the Song End Marker is set to the correct position.

2. Choose **FILE > EXPORT SONG AS AUDIO FILE**. The Export Song as Audio File dialog box will display.

Figure 9.24 The Export Song as Audio File dialog box (Mac)

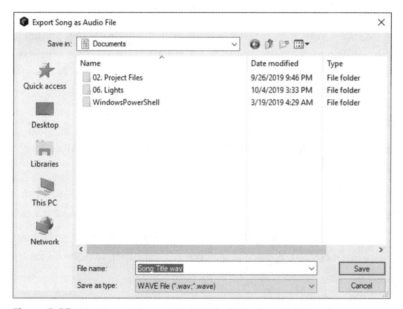

Figure 9.25 The Export Song as Audio File dialog box (Windows)

3. Specify the file name and location for your bounced file using the **SAVE AS** field (Mac) or **FILE NAME** field (Windows) and the folder navigation controls in the dialog box.

(i) By default, the bounced file will be named after the Reason song file.

4.   At the bottom of the dialog box, choose the desired file type for your bounce file. Use the **FORMAT** pop-up menu (Mac) or **SAVE AS TYPE** pop-up menu (Windows) to select Audio IFF File (aiff) or WAVE File (wav).

5.   Click the **SAVE** button to continue. Reason will display the Audio Export Settings dialog box, shown in Figure 9.22 above.

6.   Choose the desired sample rate for the bounce file from the **SAMPLE RATE** pop-up menu. Higher sampling rates will provide better audio fidelity but will also increase the size of the resulting file(s).

> **(i)** If you plan to burn your bounced audio directly to CD without further processing, choose 44.1 kHz as the sample rate for the bounce.

7.   Choose the desired bit depth for the bounced file from the **BIT DEPTH** pop-up menu.

   • Choose **16** if you plan to burn your bounce to CD without further processing.

   • Choose **24** if you want to create a final mix that will be mastered separately.

8.   Use the **DITHER** checkbox to choose whether to apply dither to the bounce file. If you choose a bit depth of 24, this checkbox will be disabled. When performing a 16-bit export, you should almost always apply dither. You would only disable the **DITHER** checkbox if you were using a plug-in device in the Master Section to apply dither instead.

9.   After confirming your settings, click the **EXPORT** button.

Monitoring Bounce Progress

When performing a bounce, Reason processes the bounce without audio playback. An Export Audio dialog box will appear, displaying the current musical bar location of the export and a progress bar indicating the overall export completion. The dialog box includes a **STOP** button that you can click to cancel the bounce.

Figure 9.26 The Export Audio dialog box

Locating the Exported File

After the bounce completes, you can retrieve your mix file from the location selected in the **EXPORT SONG AS AUDIO FILE** dialog box.

To locate your bounced mix, do the following:

1.   Using the Application Switcher, the Dock, the taskbar, or the Start menu, switch from Reason to the Finder (Mac) or File Explorer (Windows).

 The Application Switcher lets you toggle through open applications by pressing Command+Tab on a Mac or Alt+Tab on Windows.

2.   Navigate to the location you selected previously in the Export Song as Audio File dialog box.

Review/Discussion Questions

1. What are some mix attributes that are recalled with a song in Reason? What are some items that may not be recalled? (See "Recalling a Saved Mix" beginning on page 244.)

2. How can you create an automation track for a Mix Channel device? When would you need to do so? (See "Automation, Tracks, and Lanes" beginning on page 245.)

3. How can you record-enable a track for parameter automation? (See "Recording Real-Time Automation" beginning on page 247.)

4. What is the purpose of editing an automation clip? What does an automation clip look like when it is open for editing? (See "Editing Automation Points" beginning on page 251.)

5. What are some ways of editing an automation clip with the Selection Tool? (See "Using the Selection Tool for Automation" beginning on page 252.)

6. What keyboard modifier can you use with the Pencil Tool to create a horizontal automation region? (See "Using the Pencil Tool for Automation" beginning on page 253.)

7. What is the static value for an automated parameter? How can you view and change a static value? (See "Parameter Automation Static Values" beginning on page 255.)

8. What is meant by the term *mixing down*? What other term is commonly used to describe this process? (See "Creating a Mixdown" beginning on page 258.)

9. What process can you use to optimize the output levels for your mix? What device can you use in Reason for this purpose? (See "Adding a Limiter to Optimize Output Levels" beginning on page 258.)

10. What is the purpose of adding dither to a mixdown? When would you not want to use dither? (See "Adding Dither to a Mix" beginning on page 261.)

11. What are some things to consider when preparing to create a bounce of your mix? How will the bounce be affected by soloed or muted tracks? How is the end point of the bounce defined? (See "Considerations for Bouncing Audio" beginning on page 262.)

12. What sample rate and bit depth should you use when exporting a mix for use on an audio CD? (See "Selecting Export Options" beginning on page 263.)

 To review additional material from this chapter and prepare for certification, see the Reason Audio Production Basics Study Guide module available through the Elements|ED online learning platform at ElementsED.com.

Preparing the Final Mix

🎧 Activity

In this exercise, you will perform the steps necessary to create a final mix for your project in Reason. You'll begin by adding processing to the Master Section to optimize levels and prepare for exporting the mix. Then, you'll create a fade-out at the end of the song using real-time parameter automation.

🕐 Duration

This exercise should take approximately 20 minutes to complete.

⊕ Goals/Targets

- Add an MClass Maximizer device to the Master Section
- Work with parameter automation recording
- Save your work for use in Exercise 10

Exercise Media

This exercise uses media files taken from the song, "Lights," provided courtesy of Bay Area band Fotograf.

Written by: Zack Vieira and Eric Kuehnl; Performed by: Fotograf

The media provided for this course may be used for educational purposes only. No rights are granted to use the media for any other personal, commercial, or non-commercial purposes.

Getting Started

To get started, you will open your completed project from Exercise 8. This will serve as the starting point for this exercise.

Open your existing Lights project:

1. Select **FILE > OPEN** or press **COMMAND+O** (Mac) or **CTRL+O** (Windows). The Browser will open.

2. Use the Browser to navigate to the folder containing your Lights-xxx project.

3. Select your Lights-xxx project from the list of available files.

4. Click **OPEN** at the bottom of the Browser. The project will open as it was when last saved. If needed, reopen the Racks view (**WINDOW > VIEW RACKS**).

Adding Processing to the Master Section

In this part of the exercise, you will add a limiter as the only processor for the Master Section and configure the limiter to optimize the output levels for your song.

To get started, you will need to clear any existing insert effects in the Master Section.

Clear the default insert effects already loaded in the Master Section:

1. Locate the Master Section device in the upper-left area of the Racks view.

2. Click on the Master Section device to select it.

3. Choose **EDIT > CLEAR INSERT FX**.

Next, you will need to disable the Bypass button to make sure that the insert effects in the Master Section will be audible. Then you will add a limiter to the Master Section and configure it as a brickwall limiter.

Change the Master Section device so that its insert effects are not bypassed:

1. If the Master Section device is currently folded, click the **FOLD/UNFOLD** button to unfold it.

2. Click the **BYPASS MASTER INSERT FX** button on the Master Section device so that it is not lit with a blue background.

Add a limiter to the Master Section device:

1. Click the **SHOW INSERT FX** button to open the Insert FX container for the Master Section.

2. Access the Browser and navigate to the Effects location.

3. Drag and drop the **MCLASS MAXIMIZER** device to the Master Section Insert FX container.

4. Adjust the MClass Maximizer settings as follows:

    •   Input Gain: 3.0 dB

    •   4ms Look Ahead: Enabled

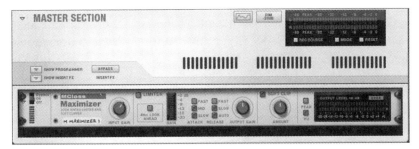

Figure 9.27 The MClass Maximizer with the suggested settings

Listen to the mix and monitor playback levels:

1. Use any technique you have learned previously to move the Song Position Pointer to the second part of the first verse, starting at Bar 33.

2. Press the **SPACEBAR** to begin playback.

3. Watch the Gain Reduction meter (**Gain**) to see where the limiter is being activated. Allow playback to continue through the first half of the verse.

4. Switch to the Main Mixer view and reset the Master Fader to its default level by **COMMAND-CLICKING** (Mac) or **CTRL-CLICKING** (Windows) on the Volume fader.

> In Exercise 7, you reduced the Master Fader by –4.0 dB to prevent clipping. That reduction is no longer needed since the MClass Maximizer device is now limiting the output.

The volume of the audio will increase, but no clip indication will appear on the meter. (The meter will display some red as the audio peaks reach the maximum possible value of 0 dBFS, but the clip indicators above the meter will not light up.)

5. Continue monitoring playback through the first chorus to observe the results during the loudest sections and verify that the mix is not clipping.

6. Stop playback when finished.

Recording Automation

The music for this song has a sudden and unnatural cutoff at the end. You can address the problem by adding a final fade-out at the end of the song. Here, you'll write a fade-out in real time using the Master Fader.

Prepare the Master Fader for automation:

1. In the Main Mixer view, **OPTION-CLICK** (Mac) or **ALT-CLICK** (Windows) on the Master Fader. Reason will display a green border around the fader, indicating that it now has an automation lane. Reason will also show the Sequencer view, which now contains an automation track for the Master Section with an automation lane for the Master Level.

2. In the Sequencer view, move the Song Position Pointer to the outro section of the song starting at Bar 93.

3. Verify that the Master Section track is ready to record parameter automation; if necessary, click the **RECORD ENABLE PARAMETER AUTOMATION** button for the track to activate it.

Figure 9.28 The Master Fader in the Main Mixer view with a green automation border

Figure 9.29 The Master Section automation track with an automation lane for the Master Level; Record Enable Parameter Automation is active.

Automate a fade-out on the Master Fader track:

1. Verify that the Song Position Pointer is at Bar 93 and click the **RECORD** button in the Transport Panel to begin a record take.

2. Click on the Master Fader in the main mixer and slowly move the fader all the way down to minus infinity (−∞). You'll have about 10 to 15 seconds to complete the fade-out.

3. Keep an eye on the Position/Time placements in the Transport Panel and the clips in the sequencer during your fade-out. Be sure to complete the fade-out before you reach the end of the clips at Bar 101. Once you have faded all the way down, you can release the fader and it will stay in position.

4. Stop recording once automation has been written a few bars past the end of the song.

5. Return the Song Position Pointer to Bar 93 and press the **SPACEBAR** to listen to the fade-out and verify the automation.

6. If you're not satisfied with the result, try another pass of writing the automation. It may take several attempts to get the timing of the fade-out just right!

(i) If you want to remove automation you've recorded, choose EDIT > UNDO RECORD TRACK, or press COMMAND+Z (Mac) or CTRL+Z (Windows). Use the undo command immediately after stopping recording, before making other changes.

View the volume automation on the Master Section track:

1. Choose **WINDOW > VIEW SEQUENCER** or press **F7** to display the Sequencer view and make it active.

2. If needed, scroll the **Master Section** track into view. The automation lane will contain a clip with the results of the fade-out you wrote for the track.

   You can double-click the clip with the Selection Tool to open it and view the automation points.

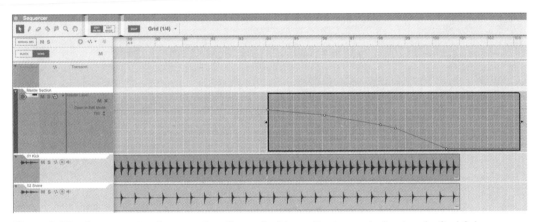

Figure 9.30 The Master Level automation clip on the Master Section track showing the final fadeout

Finishing Up

To complete this exercise, you will need to save your work and close the project. You will be reusing this project in Exercise 10, so it's important to save the work you've done.

Finish your work:

1.   Press **[.]** (**PERIOD**) on the numeric keypad to move the Song Position Pointer to the beginning of the song.

2.   Choose **FILE > SAVE** to save the project.

3.   Choose **FILE > CLOSE** to close the project.

That completes this exercise.

Chapter 10

Beyond the Basics

...What to Explore to Become a Power User...

In this chapter, we take a brief look at various advanced features in Reason. We begin by delving into the use of cabling on the back of the rack to create custom audio and CV routing. Then we explore some fundamental MIDI editing techniques that can be used to create and modify MIDI data. Next we discuss how you can simplify project management with submixing. The chapter concludes with a sample of some keyboard shortcuts that you can use to dramatically accelerate your editing workflow.

✦ Learning Targets for This Chapter

- Understand the details of rack cables

- Explore fundamental MIDI editing techniques

- Gain familiarity with submixing techniques

- Learn essential keyboard shortcuts

Key topics from this chapter are illustrated in the Reason Audio Production Basics Study Guide module available through the Elements|ED online learning platform. Sign up at ElementsED.com.

For those who are new to the world of DAWs, or simply new to Reason, the information required to produce a finished project can seem a bit overwhelming. Learning the fundamentals should obviously be the first goal, so that you'll have a strong foundation. But it's also helpful to learn some advanced features so you'll be able to work quickly and effectively.

The goal of this chapter is to present some key features in Reason that you can start using right away. Many of the topics we touch on here are easily deep enough to warrant an entire chapter in an advanced course. However, here we address just the basic functionality so that you'll be able to put these features to use quickly. Once you've gained a bit of experience with Reason, you can come back to these topics at a later date and learn them comprehensively.

Cable Routing in the Rack

Reason connects devices in the rack together with virtual cables that route signals. In many cases, the flexible automatic routing features in Reason allow you to work with devices without having to directly manipulate the cable connections. As you create devices, the automatic routing works to build a logical signal flow based on the types of devices and the positioning of devices in the rack.

However, many powerful sound design opportunities arise from the ability to configure your own signal routing. Also, understanding how to work with cabling in the rack can be very helpful in situations where you need to troubleshoot the signal flow if you aren't getting the results you expect.

 Use the Tab key to switch between the front and back views of the rack. You can also choose Options > Toggle Rack Front/Rear to change the view.

Reason supports two types of connections on the back of the rack: audio signal connections and CV signal connections.

 CV stands for Control Voltage. A CV signal in Reason is a signal from one device that can be used to control or manipulate parameters in another device.

Audio Signals and Cables

Audio signal routing is the most common type of routing that you will encounter on the back of the Reason rack. Each audio cable represents a connection that transmits an audio signal between two devices. Audio cables link the audio output jacks of one device to audio input jacks of another device.

Input and output jacks have the same appearance, but labels are provided to distinguish input jacks from output jacks. Reason will prevent you from making a cabling connection between two jacks of the same type (input or output).

In Figure 10.1, an audio signal is generated by the Thor synthesizer instrument. The audio output jacks of the Thor instrument connect to the audio input jacks of the Scream 4 effect device. The Scream 4 device applies an audio effect and routes the signal to its output jacks, which then connect to the input jacks of a Mix Channel device.

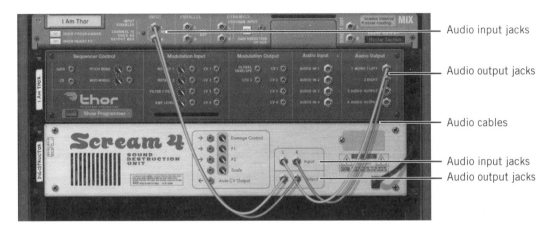

Figure 10.1 Audio cables and jacks routing audio from Thor through Scream 4 and then to a Mix Channel device

Despite the fact that the virtual rack cables mimic analog patch cables in appearance, there is no analog-to-digital or digital-to-analog audio conversion happening between devices. The cables create digital connections inside the software. Audio is converted between the analog and digital domains only when it passes through your audio interface. In Reason, these conversions are represented by connections to the Hardware Interface device and connections to Audio Track devices that receive input from your audio interface.

CV Signals and Cables

The abbreviation CV stands for Control Voltage, a term borrowed from legacy electronic devices. In the pre-MIDI days of synthesizers, described in Chapter 4, CV signals were created to control various synthesizer components. For example, CV signals might be used to open and close a filter, turn an amplifier up and down, set the pitch of an oscillator, and so on. It might seem odd at first to have a concept from the early days of electronic music represented in a modern DAW like Reason, but the CV features in Reason allow for unique sound manipulation possibilities.

CV signals are most commonly used to modulate a parameter on a device according to values from a modulation source. A common modulation source is a low-frequency oscillator, or LFO. As the name suggests, an LFO is a wave that cycles in amplitude at a relatively slow rate; LFO frequencies are typically well below the 20 Hz threshold of human hearing, so you would not be able to hear an LFO if you played it back as an audio signal.

Instead, an LFO can be used to change a parameter on a device such as a synthesizer in a cyclical pattern. For example, you could connect an LFO CV output to the amplifier CV input of an instrument to automatically move the volume of the instrument up and down, creating a tremolo effect.

Figure 10.2 shows an example of CV routing used to create a tremolo effect. A CV output jack from a Pulsar Dual LFO device is linked to the **Master Volume** CV input jack on an NN-XT sampler instrument.

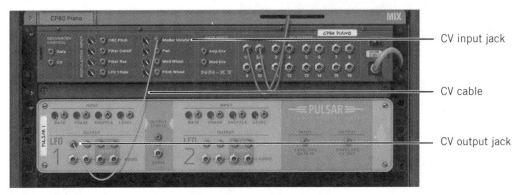

Figure 10.2 CV cables and jacks used for a tremolo effect

The LFO signal will cause the volume of the instrument to increase and decrease. The rate of the volume change (LFO frequency) and the intensity of the volume change (LFO amplitude) can be modified with the Rate and Level controls on the front panel of the Pulsar device.

CV jacks on devices have a smaller visual appearance than audio jacks. As with audio jacks, CV jacks on a device are either output jacks or input jacks, denoted by labels on the device. Reason won't let you connect two CV output jacks or two CV input jacks together.

In addition to modulation signals such as LFOs, CV connections can also transmit gate and note signals. A gate CV signal tells an instrument when to turn a sound on or off. A note CV signal tells an instrument the pitch of a note that the instrument should play. Gate and note signals are used with devices such as the RPG-8 Monophonic Arpeggiator and the Matrix Pattern Sequencer to allow those devices to control other instruments in the rack. Player devices can be used for similar purposes.

Cable Color Coding

Reason represents different types of cable connections with different colors:

- **Red**—Red cables represent audio connections to "mixer" devices such as Mix Channel devices or Line Mixer 6:2 devices.

- **Green**—Green cables represent audio connections to effects devices.

- **Blue**—Blue cables represent connections to Combinator utility devices.

- **Yellow**—Yellow cables represent CV connections. CV cables also have a slightly thinner appearance than audio cables.

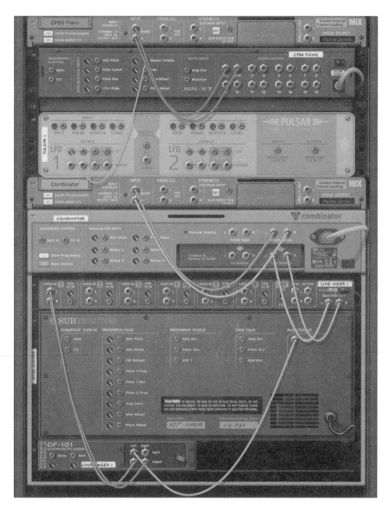

Figure 10.3 Color-coded cables indicating different connection types

Connecting and Disconnecting Cables

Cable connections in the rack can easily be changed using drag-and-drop operations.

To create a connection between jacks in the rack:

1. Click and hold on an empty jack that is not yet connected. A hanging cable will appear, and the other connections on the screen will dim.

2. Drag the cable to a compatible destination jack. To connect from an audio output jack, for example, drag to an audio input jack.

When you hover over a valid destination, the target jack will light up with a red color.

3. Release the mouse over a valid destination to drop the cable and complete the connection.

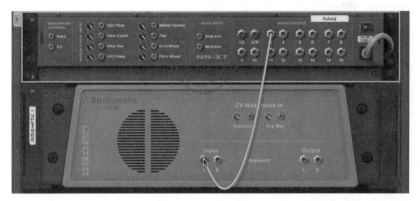

Figure 10.4 Dragging an audio cable from an NN-XT device output jack to an Audiomatic device input jack

(i) You can drag from an output to an input or from an input to an output. As long as the source and destination are compatible, the starting point doesn't matter.

To move an existing connection to a different jack:

1. Click and hold on a jack that has an existing cable connection. The cable will highlight and the other connections on the screen will dim.

2. Drag the connector to another compatible destination jack.

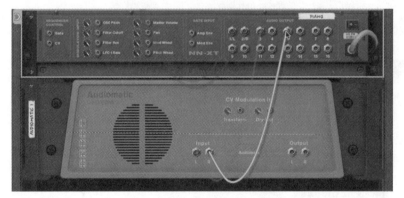

Figure 10.5 Moving a cable connection from one NN-XT device output jack to another

3. Release the mouse to drop the cable and complete the connection.

To disconnect an existing connection:

1. Click and hold on a jack that has an existing cable connection. The cable will highlight and the other connections on the screen will dim.

2. Drag the connection away from the jack so that it is not positioned over a device jack.

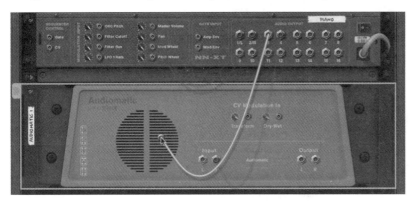

Figure 10.6 Dragging a cable away from a jack to disconnect it

3. Release the mouse button to drop the cable and remove the connection.

You can also connect and disconnect cables using pop-up menus in the rack.

To connect a cable using a menu:

1. Right-click on the jack you want to connect. Reason will display a menu listing all available devices in the rack.

2. Mouse over a destination device. A submenu will appear, listing all of the valid destination jacks. (See Figure 10.7.)

3. Select the desired destination jack to complete the connection.

Any destination jacks marked with an asterisk in the menu have an existing cable connection. If you select one of these destinations, the existing cable connection will be replaced.

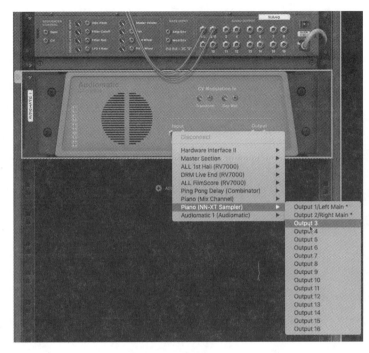

Figure 10.7 Creating a cable connection using a menu

To disconnect a cable using a menu:

1. Right-click on a jack with an existing connection that you want to disconnect.

2. Choose **DISCONNECT** from the pop-up menu.

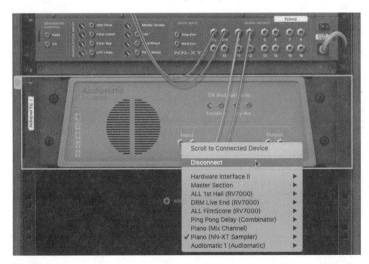

Figure 10.8 Disconnecting a cable connection using a menu

Reduce Cable Clutter

Typical productions in Reason often use a large number of devices. As you create more and more devices, the number of cables crisscrossing the back view of the rack can become overwhelming. The view is especially complex when many rack devices are unfolded.

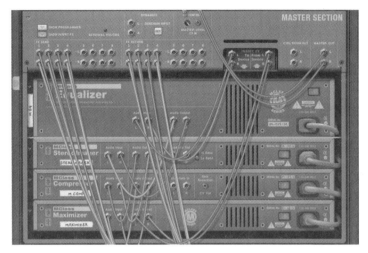

Figure 10.9 Overlapping cable connections for insert effects and send effects on the Master Section

In these situations, it can be hard to see where the connected cables are going. The Reduce Cable Clutter option can help.

To toggle this option on and off, choose **OPTIONS > REDUCE CABLE CLUTTER** or press the **K** key. By default, activating Reduce Cable Clutter makes all the cables in the track translucent except for the cables connected to the selected device, which remain opaque.

Figure 10.10 Cable connections displayed with Reduce Cable Clutter activated

Reason preferences allow you to choose one of three settings that determine the way the Reduce Cable Clutter option works. The behavior described above uses the default setting: Shows Cables for Selected Devices Only.

Mono and Stereo Audio Connections

Almost all devices in Reason that support audio cable connections provide stereo sets of jacks. For example, the MClass Equalizer device seen in Figure 10.10 contains two jacks labeled Audio Input. The jack labeled L receives the left channel of a stereo signal, while the jack labeled R receives the right channel. To receive a complete stereo signal, both jacks must be connected.

The Subtractor synthesizer, on the other hand, is an example of a *monaural* device that can only produce a single channel of audio information. It has only a single audio output jack.

Figure 10.11 A Subtractor device with a single audio output connection to a Mix Channel device

Although all effect devices have stereo pairs of input and output jacks, it is generally possible to use these devices with mono signals. For example, to use the MClass Equalizer device with a mono signal, you can connect a single cable to the L audio input jack. When used alone, this jack acts as a mono input. Likewise, connecting a single cable from the L audio output jack will cause it to act as a mono output.

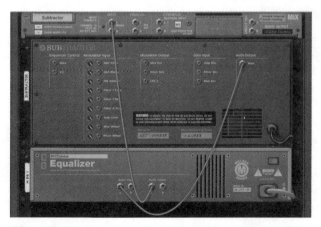

Figure 10.12 A mono signal path through an MClass Equalizer device; the R input and output jacks remain unconnected

Different effects devices support different combinations of input and output cables. For example, the RV7000 MkII Reverb device can receive a mono input signal but produce a stereo reverb output signal. To indicate which effect device input and output jacks can be connected, devices have signal flow graphs on their back panels.

Figure 10.13 Signal flow graphs on the MClass Equalizer device indicate support for both mono and dual mono processing.

Five different types of signal flow graphs can appear on an effect device, as shown in the following table.

Signal Flow Graph	Graph Translation
	The effect device supports a purely mono signal path; you can connect just the L input and L output to process a mono signal.
	The effect devices can produce stereo output from mono input. You can connect just the L input to send in a mono signal but connect both the L and R output jacks to send out a stereo result.
	The effect device performs dual mono processing. You can connect the L and R input and output jacks in a full stereo configuration. Internally, the device will apply its effect to the left and right channels separately.
	The effect device performs summed mono processing. You can connect the L and R input and output jacks in a full stereo configuration. Internally, the device will sum the left and right input channels to a mono signal and then apply an effect, which produces a stereo output result.
	The effect device performs true stereo processing. You can connect the L and R input and output jacks in a full stereo configuration. Internally, the device will use both the left and right input channels to produce unique information for the left and right output channels.

Whenever Reason performs automatic cable routing, such as when you create a new device or move a device with the Shift modifier, the automatic routing process will connect the device in a stereo configuration if possible. Similarly, when you manually connect and disconnect cables on the back of the rack, Reason assists in configuring stereo signals. If you connect the L jack of a stereo pair to the L jack of another stereo pair, Reason will automatically connect the R jacks for you. Likewise, if you disconnect the L jack for a stereo pair of cables, Reason automatically disconnects the R jack as well.

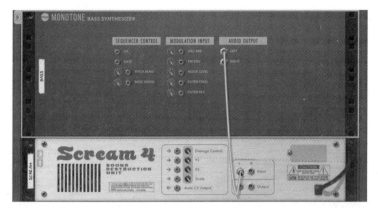

Figure 10.14 After dragging the left output of the Monotone device to the Scream 4 device...

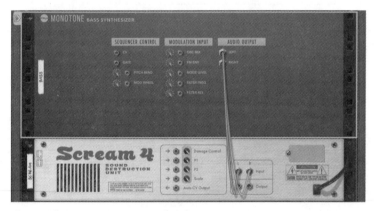

Figure 10.15 ...Reason automatically connects the right output as well.

Routing Without Cables

Not all signals in Reason Intro are routed using cables in the rack. Some signal paths are handled behind the scenes in the software. It can be helpful to recognize a few of these cases to avoid confusion when working in the rack.

- **Instrument Tracks and Instrument Devices:** Instrument devices have a direct internal connection to their corresponding instrument tracks through which the instrument devices receive MIDI performance information. The rack has no MIDI cables representing this connection.

- **Audio Track Input:** Audio Track devices do not have cable input jacks on their back panels. Instead, Reason routes audio input to an audio track directly from the input chosen with the Audio Input drop-down list. No cable connection is involved with this routing.

- **Mix Channel and Audio Track Mixer Routing:** Audio Track and Mix Channel devices don't use cables to send audio into the main mixer. Instead, an internal connection that Reason calls a P-LAN connection routes these device signals into the main mixer.

Mix Channel and Audio Track devices do have a set of Direct Out jacks that can route their signals to a destination other than the main mixer in advanced scenarios. These jacks are usually disconnected.

MIDI Editing Techniques

MIDI editing is another topic worthy of its own chapter (or several chapters). But as with cable routing, a little knowledge is all you need to get started. Understanding the basic MIDI editing is absolutely essential in the modern studio. Whether your goal is to add a few MIDI instruments to a song or to compose a virtual orchestral masterpiece, you'll need to know how to edit MIDI data quickly and accurately.

Sequencer Edit Mode

MIDI editing begins in the Sequencer view with clips on note lanes. Clips are often created by recording a MIDI performance from a controller, such as a MIDI keyboard. However, you can also create empty note clips by single-clicking or clicking and dragging with the Pencil Tool in a note lane. Double-clicking a note lane with the Selection Tool will also create an empty clip.

Editing the contents of a note clip requires switching the Sequencer view to Edit Mode. Edit Mode replaces the Arrangement pane with a dedicated panel for viewing and editing MIDI data across a single track or multiple tracks.

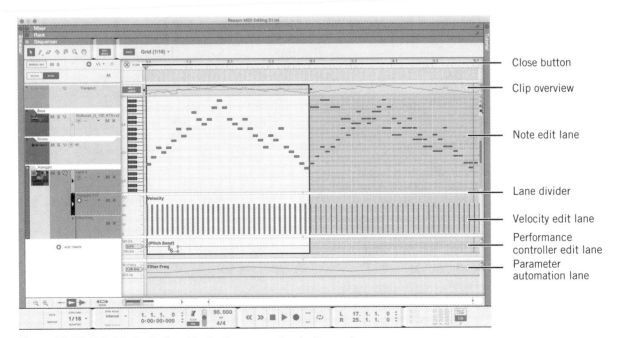

Figure 10.16 Edit Mode displayed in the Sequencer view in Reason Intro

Opening Edit Mode

Reason provides many ways to open Edit Mode.

To open Edit Mode, do one of the following:

- Double-click a note clip with the Selection Tool.

- Click on a note clip to select it and press **RETURN** (Mac) or **ENTER** (Windows).

- Right-click a note clip and choose **EDIT** from the pop-up menu.

- Click on a note clip to select it and choose **EDIT > EDIT** from the main menus.

- Click on a note clip to select it and press **COMMAND+E** (Mac) or **CTRL+E** (Windows).

- Click on a note clip to select it and then click the **EDIT MODE** button in the Sequencer toolbar.

Figure 10.17 The Edit Mode button in the Sequencer toolbar

- Click on a note clip to select it and then press **SHIFT+TAB** to switch to Edit Mode.

> (i) Although you can open Edit Mode without first selecting a clip in the Arrangement pane, starting with a selected clip is often easier and more efficient.

Focusing Lanes and Tracks in Edit Mode

Once Edit Mode is visible, it will automatically open the clip that was selected in the Arrangement pane so you can begin making changes. From within Edit Mode, though, you can switch between different clips and note lanes that you want to edit.

To switch edit focus between clips on a lane:

1. Identify a closed clip you would like to open. Closed clips have a grayed-out/faded appearance in the Note Edit Lane.

2. Using the Selection tool, double-click the closed clip in the Note Edit Lane or the Clip Overview. The clip will become highlighted and the notes will appear colorized.

Edit Mode provides two different approaches to focusing lanes and tracks, depending on whether the Multi Lanes button to the left of the Clip Overview area is active.

Figure 10.18 The Multi Lanes button in Edit Mode

When Multi Lanes mode is inactive, Edit Mode will display the contents of only a single note lane at a time. To switch focus between different note lanes, click on the Note Lane Handle in the Track List for the desired lane. Clicking a Note Lane Handle will also select the associated track in the Track List.

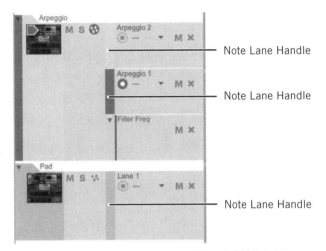

Figure 10.19 Note Lane Handles in the Track List

When Multi Lanes mode is active, Edit Mode can display the contents of multiple lanes simultaneously. You will still be limited to making changes on one lane at a time, but seeing the contents of other lanes can help you edit the notes so that they fit together with the other parts of your song.

To use Multi Lanes mode, first click the **MULTI LANES** button to activate it. Then, select multiple tracks in the Track List corresponding to the lanes you want to view.

To select multiple tracks, do one of the following:

■ Hold **COMMAND** (Mac) or **CTRL** (Windows) and click on a track in the Track List to add it to the set of selected tracks.

■ With one track initially selected, hold **SHIFT** and click on another track. Doing so will select the second track as well as all the tracks between the first and second track.

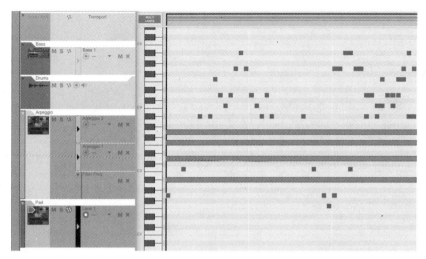

Figure 10.20 Edit Mode displays notes from multiple tracks with Multi Lanes mode active.

With multiple tracks selected and Multi Lanes mode active, Edit Mode will display the notes from all of the lanes on the selected tracks at the same time. As shown in Figure 10.20, only one of the displayed lanes can have edit focus at a time. The notes for that lane will appear in different shades of red, while notes from other lanes will be grayed out in the background.

To switch edit focus to a different visible lane, do one of the following:

- Click on a Note Lane Handle in the Track List.

- Click on one of the gray background notes. Edit focus will switch to the lane that contains the note.

In Multi Lanes mode, all of the Note Lane Handles in the Track List display arrow icons. The Note Lane Handle for the lane with edit focus will appear active with a dark background. The Note Lane Handles for lanes of other selected tracks will have gray backgrounds, while the Note Lane Handles for tracks that are not selected will have a faded appearance.

(i) If Multi Lanes mode was active when Edit Mode was last closed, you can select multiple tracks or clips in the Sequencer prior to opening Edit Mode to start with all of the selected contents already visible.

(i) To resize lanes in Edit Mode, click and drag the dividers separating the lanes.

Arranging Versus Editing Grid Settings

Reason maintains separate settings for the grid size when working in the Arrangement pane and when working in Edit Mode. In the Arrangement pane you will usually set the grid size to a relatively large value, such as **BAR** or **1/2,** so that you can move clips quickly across the grid. When you open a clip in Edit Mode, you can set a second grid size value. Here, you will often use a smaller unit, such as **1/16,** so that you can edit notes with precision.

Reason will switch to the primary arrangement grid size when you click in the Clip Overview area of the Edit Mode interface. In the Clip Overview area, you can perform clip-arrange operations as in the main Arrangement pane. Nonetheless, having the grid size change in Edit Mode can be unexpected. If the grid value doesn't match what you are expecting, use the Selection Tool and click either the Clip Overview area or the Note Edit Lane to make sure edit focus is set to the correct location.

Closing Edit Mode

When you finish working in Edit Mode, you can return to the Arrangement pane. To close Edit Mode, do one of the following:

■   Press **ESCAPE**.

■   Click the **CLOSE** button in the Edit Mode panel.

■   Click the **EDIT MODE** button in the Sequencer toolbar.

■   Press **COMMAND+E** (Mac) or **CTRL+E** (Windows).

■   Press **SHIFT+TAB** to switch back to the Arrangement pane.

Editing MIDI Data

Reason Intro offers the same powerful MIDI editing toolset that you'll find in the full version of Reason and Reason Suite. These tools can be used to create and edit MIDI performances, regardless of whether you own a MIDI controller, such as a keyboard.

Let's take a look at some options for creating and working with MIDI data in Reason.

Creating Notes

Manually entering MIDI note data may seem like a slow method of creating a performance, compared to playing the performance on a MIDI controller. But in fact, many music producers prefer to manually enter note data for certain types of tracks (such as drums, percussion, and bass). With a little practice, you can manually create MIDI note data very quickly and accurately.

To manually create a note:

1.   Open a note clip in Edit Mode.

2.   If necessary, click on the clip with the Selection Tool to ensure that the editing grid settings are active.

3.   Enable the **SNAP** function if it is not already enabled.

4.   Set the grid size value to the desired rhythmic interval. (A 1/16-note interval is commonly used when programming drums and musical instruments.)

5.   Select the Pencil Tool.

6.   Click and hold with the Pencil Tool at the location (in both pitch and time) where you would like to enter a note. You can adjust the note's duration by dragging forward/backward before releasing the mouse button. (See Figure 10.21.)

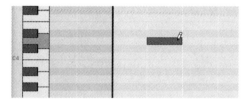

Figure 10.21   Manually entering a note with the Pencil Tool

(i)   If the Snap function is not active, adding a note with the Pencil Tool will create a 128th note. The note starting position will not snap to the grid.

To manually create a sequence of notes at the same pitch:

1.   Follow Steps 1 through 5 above.

2.   With the Pencil Tool active, click on the Pencil Tool icon in the toolbar to display the associated pop-up menu. Two different modes for the Pencil Tool will be available.

3.   Select the **DRAW MULTIPLE NOTES** mode from the pop-up menu.

4.   Click and hold with the Pencil at the location where you would like the sequence of notes to begin. Drag the Pencil to the right to create a string of notes at the same pitch.

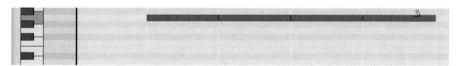

Figure 10.22   Manually entering a sequence of notes at the same pitch

5.   Once you have created the desired number of notes, release the mouse button.

Selecting Notes

Another essential part of working with MIDI data involves selecting notes. Understanding how to quickly select single notes and groups of notes is critical to the MIDI editing workflow. You can use the following techniques to select notes in a clip in Edit Mode.

To select an individual note, do the following:

- Using the Selection Tool, click anywhere on a note.

To select groups of notes, do one of the following:

- Using the Selection Tool, **SHIFT-CLICK** (Mac) or **CTRL-CLICK** (Windows) on each of the desired notes.

- Using the Selection Tool, click an empty area in the clip and draw a rectangle around the target notes.

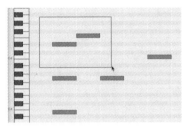

Figure 10.23 Selecting a group of notes with the Selection Tool

- Choose **EDIT > SELECT ALL** or press **COMMAND+A** (Mac) or **CTRL+A** (Windows) to select all notes in the clip.

Editing Notes

With basic note creation and selection techniques in hand, a whole range of MIDI editing functions are available to you. Let's look at some essential note editing techniques that you'll need to know.

Deleting Notes

To delete an individual note, do one of the following:

- Using the Selection Tool, double-click on a note.

- Using the Erase Tool, click on a note.

To delete a group of notes:

- Select the notes and press **DELETE** (Mac) or **BACKSPACE** (Windows).

- Using the Erase Tool, click an empty area in the clip and draw a rectangle around the target notes.

Muting Notes

Rather than deleting notes, you can also mute them to prevent them from playing back.

To mute an individual note:

■   Using the Mute Tool, click on a note.

To mute a group of notes:

1.   Select the target notes.

2.   Press the **M** key.

Apply the same process on notes that are already muted to un-mute the notes.

Moving Notes

To move a note to a new location:

■   Using the Selection Tool, click on the note and drag left or right.

To move a group of notes to a new location, do the following:

1.   Select the target notes.

2.   Using the Selection Tool, drag any one of the notes left or right. All selected notes will move together.

(i)   With the Snap function disabled, notes can be moved freely. When the Snap function is enabled, notes move in steps according to the selected grid size.

Transposing Notes

To change a note's pitch:

■   Using the Selection Tool, click on the note and drag up or down.

To change the pitch of a group of notes, do the following:

1.   Select the target notes.

2.   Using the Selection Tool, drag one of the notes up or down. All selected notes will transpose together.

   When moving or transposing notes, hold Option (Mac) or Ctrl (Windows) to copy the notes. This lets you quickly create harmonies or chords for melodic instruments.

Resizing Notes

To resize a note, do the following:

1. Click on the note with the Selection Tool to select the note.

2. Click and drag one of the resize handles that appear on the left and right sides of the selected note.

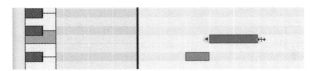

Figure 10.24 Resizing a note with the Selection Tool

To resize a group of notes, do the following:

1. Select the target notes.

2. Click and drag on a resize handle for any selected note to change the start or end point of all the selected notes.

Working with Velocity

Most modern MIDI controllers are *velocity sensitive*, meaning that they automatically detect how hard a key or pad is struck. The velocity data is transmitted and recorded as a characteristic of the MIDI note. However, when notes are manually entered (with the Pencil Tool, for example), they will all share a default velocity value. This can lead to very stiff, robotic-sounding performances. Modifying and manipulating velocity data is essential to creating great-sounding MIDI parts.

Note velocity is displayed in the Velocity Edit Lane in Edit Mode. Velocity values are displayed using vertical indicators known as *velocity bars*. You can edit velocity values by raising or lowering the velocity bars for individual notes.

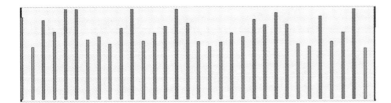

Figure 10.25 A portion of the Velocity Edit Lane in Edit Mode

To edit MIDI velocity data for a single note, do one of the following:

- Use the Pencil Tool to click and drag the associated velocity bar up or down.

- Use the Pencil Tool and single-click at the desired height to set the bar to that specific position.

To smoothly change the velocity of multiple notes:

1. Select the Pencil Tool.

2. Place the Pencil Tool in the Velocity Edit Lane and hold **OPTION** (Mac) or **CTRL** (Windows). The pencil icon will change to a cross shape.

3. Click and drag to draw a line in the Velocity Edit Lane.

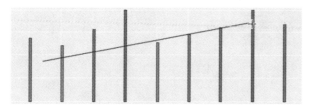

Figure 10.26 Drawing a velocity line in a Velocity Edit Lane

4. Release the mouse button to adjust the velocity bars to match the height of the line. You can use this technique to draw velocity ramps or set multiple notes to a similar velocity.

In situations where a clip contains a large number of notes or a cluster of overlapping notes, changing the velocity bars for just a few of the notes can be difficult. In these cases, first select the notes corresponding to the velocity bars you want to change. Then, hold the **SHIFT** modifier as you make velocity changes with the Pencil Tool. The change will apply to the velocity bars for the selected notes only.

The Tool Window

During the music production process, you'll frequently want to make adjustments to MIDI data. Reason provides a window called the Tool Window, which allows you to transpose, quantize, and otherwise fine-tune your MIDI performances. This window gives you many choices and extensive control over the changes you apply to MIDI data. To access the Tool Window, choose **WINDOW > SHOW TOOL WINDOW** or press Function Key **F8**.

The Tool Window contains two tabs: the Sequencer Tools tab and the Groove Settings tab. Here we'll take a quick look at the Sequencer Tools tab.

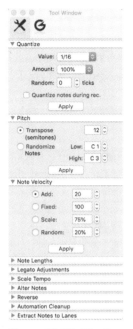

Figure 10.27 The Sequencer Tools tab of the Tool Window displaying commonly used functions

The Tool Window can be used to quickly apply the following modifications to selected notes or clips on an instrument track:

- **Quantize**—Clean up the timing of MIDI performances by moving notes closer to the grid.

- **Pitch**—Transpose notes up or down by a specified number of semitones.

- **Note Velocity**—Scale the velocity of notes up or down, set velocities to a fixed value, or randomize velocities.

- **Note Lengths**—Shorten or lengthen notes or set them equal to a specific length.

To use a function in the Tool Window:

1. Select the note(s) or clip(s) you want to modify.

2. Configure the parameters for a function in the Tool Window.

3. Click the **APPLY** button for the function in the Tool Window.

For more information on Tool Window functions, consult the documentation for your Reason software. Exercise 10 contains an example of using the Pitch function of the Tool Window.

Submixing

Submixing is a technique whereby multiple mixer channels are summed to a common destination channel to simplify certain mixing tasks. While submixing is most useful in sessions with a large number of tracks, there are advantages to using submixes in smaller projects as well.

Simplifying a Mix

Submixing is frequently used to help simplify the mixing process. By submixing a group of related tracks, you can more easily apply effects across the whole group. You can also automate the summed levels of the group using a single fader.

Submixing is an essential technique for managing drum recordings, which typically span a number of tracks. By routing the drums to a common destination, you can apply compression or limiting to the group (for example) using a single device. Then, you can quickly set the level for the entire drum kit using a single fader, while maintaining the relative levels of the separate tracks, as set by their individual faders.

But submixing isn't just for drums. It's a great way to manage any related group of tracks, such as background vocals, guitars, and keyboards.

To create a submix, follow these steps:

1. In the Main Mixer view, select all of the mixer channels that you wish to include in the submix.

2. Press **COMMAND+G** (Mac) or **CTRL+G** (Windows) or choose **EDIT > ROUTE TO > NEW OUTPUT BUS**. Reason will create a new Mix Channel device and route the outputs of the selected mixer channels to the new bus channel.

Figure 10.28 A drum submix with the source channels (purple) routed to a bus channel (blue with a red fader)

Figure 10.29 The output bus channel has a corresponding Mix Channel device in the Racks view.

Once you've routed the source tracks to a submix destination, you can easily insert effect devices on the output bus Mix Channel device to have them apply to the submix. You can also adjust the volume and panning on the output bus channel to modify the submix level and position in the larger mix.

Creating Stems

Another important use for submixes is to aid in the creation of *stems*. Stems are mixed-down versions of your submixes that can be useful when mixing or remixing a project in a different DAW. Stems can also be used to supplement a live performance. If you've routed all of your tracks to submixes, it is very easy to create stems in Reason.

To create stems from submixed tracks, do the following:

1. Move the left and right Loop Locators to span the section of the song you would like to export as stems. This might be the entire song, or it might be one section of the song, such as a verse or chorus.

2. Locate the first submix output bus channel that you would like to export as a stem.

3. Click the **SOLO** button on the submix output bus channel. All of the tracks in the submix will automatically be soloed.

4. Play the session to verify that no other tracks are currently audible. You should still be able to hear send effects, such as reverbs and delays, applied to the submixed tracks.

5. If you are using any devices on the Master Section (such as a limiter), bypass or remove them.

    (i) Generally when using submix tracks in another DAW, you will want to apply master bus processing there for maximum control. However, you can also create stems with processing in place on the Master Section to help preserve the sound of your mix.

6. Choose **FILE > EXPORT LOOP AS AUDIO FILE**. The Export Loop as Audio File dialog box will appear.

7. Choose the desired export settings as described in Chapter 9. The Export Loop as Audio File dialog works just like the Export Song as Audio File dialog. After completing the export process, the submix will be available as an audio file.

8. Repeat the process for each of the submixes in your project.

When you are finished, you will have an audio stem file for each of the submixes. These files can then be aligned in another Reason project or in a different DAW. Using these stems, you can re-create the original mix with no more than a few small tweaks.

The Bounce Mixer Channels Function

Reason Intro includes a feature called Bounce Mixer Channels (File > Bounce Mixer Channels) that allows you to export multiple submixes at once to create multiple stems in one step. However, when using Bounce Mixer Channels, send effects must be exported as separate files rather than being included in the stems themselves. Also, any Master Section effect devices that you intended to apply to the stems will not be included.

Soloing a submix and using the Export Loop as Audio File command ensures that the exported stem file exactly matches what you hear when you play back the submix. Take some time to experiment with the Bounce Mixer Channels feature, though, as it is very useful in many situations!

Keyboard Shortcuts

In closing, we'll cover some keyboard shortcuts that will dramatically accelerate your work in Reason. Learning even a handful of essential shortcuts can elevate your status from prospect to pro user in no time.

Function Keys

The following tables show uses for the Function keys in Reason. These are the keys labeled **F1** through **F12** at the top of a standard alphanumeric keyboard.

Main Views

Key	Function
F5	Main Mixer View
F6	Racks View
F7	Sequencer View

Other Views and Windows

Key	Function
F2	Spectrum EQ Window
F3	Browser
F8	Tool Window

Sequencer Shortcuts

The following tables show uses for single-key shortcuts that activate commands in the Sequencer view.

Sequencer Tools

Key	Function
Q	Selection Tool
W	Pencil Tool
E	Eraser Tool
R	Razor Tool
T	Mute Tool
Y	Magnifying Glass Tool
U	Hand Tool
I	Speaker Tool (when available)

Editing Functions

Key	Function
S	Enable/Disable Snap Function
M	Mute/Un-mute Selected Notes or Clips
RETURN (Mac) / ENTER (Windows)	Open a Selected Clip in Edit Mode
ESCAPE	Close Edit Mode
X	Enable/Disable Crossfade on Selected Clips

Rack Commands

The following table shows keyboard shortcuts that can be useful when working in the Racks view.

Key	Function
TAB	Switch Between Front/Back View of Rack
K	Enable/Disable Reduce Cable Clutter
COMMAND/CTRL+T (Mac/Windows)	Create Audio Track
COMMAND/CTRL+SHIFT+M (Mac/Windows)	Create Mix Channel Device

Transport Commands

The following table shows keyboard shortcuts that can be used to control functions in the Transport Panel.

Key	Function
C	Enable/Disable Metronome Click
L	Enable/Disable Loop Mode
[.] (Period on numeric keypad)	Navigate to Beginning of Song
COMMAND+RETURN (Mac) CTRL+ENTER (Windows)	Record
OPTION/ALT+LEFT ARROW (Mac/Windows)	Navigate to Left Loop Locator
OPTION/ALT+RIGHT ARROW (Mac/Windows)	Navigate to Right Loop Locator

Review/Discussion Questions

1. What types of signals can be routed with cables on the back of the Reason rack? (See "Cable Routing in the Rack" beginning on page 276.)

2. What does CV stand for? How are CV signals used in Reason? (See "CV Signals and Cables" beginning on page 277.)

3. What are some ways to connect and disconnect cables on the back of the rack? (See "Connecting and Disconnecting Cables" beginning on page 279.)

4. What is the effect of the Reduce Cable Clutter option? How can you turn this option on and off? (See "Reduce Cable Clutter" beginning on page 283.)

5. Why do most devices have left and right pairs of audio jacks? Give an example of a situation where only the left jack of a pair would be connected. (See "Mono and Stereo Audio Connections" beginning on page 284.)

6. What are some signals in Reason that are *not* routed by cable connections in the rack? (See "Routing Without Cables" beginning on page 286.)

7. What is Edit Mode in the Sequencer view? How can you open a clip in Edit Mode? (See "Sequencer Edit Mode" beginning on page 287.)

8. How can you switch between tracks and lanes in Edit Mode? What does the Multi Lanes button do? (See "Focusing Lanes and Tracks in Edit Mode" beginning on page 288.)

9. How can new notes be manually created in a clip? (See "Creating Notes" beginning on page 291.)

10. What are some common techniques for editing notes? (See "Editing Notes" beginning on page 293.)

11. What is MIDI velocity? How is velocity data viewed and edited in Reason? (See "Working with Velocity" beginning on page 295.)

12. What are some functions provided by the Tool Window? (See "The Tool Window" beginning on page 296.)

13. What parts of the music production process can be simplified using submixes? (See "Submixing" beginning on page 298.)

14. Why is it helpful to create stems when exporting your Reason Intro project to be imported into a different DAW? (See "Creating Stems" beginning on page 299.)

 To review additional material from this chapter and prepare for certification, see the Reason Audio Production Basics Study Guide module available through the Elements|ED online learning platform at ElementsED.com.

Finalizing a Project

🎧 Activity

In this exercise, you will perform the final steps necessary to finish a project in Reason. You'll begin by creating a drums submix. Then, you'll duplicate and transpose the **Strings** track to achieve a bigger string sound. Finally, you'll export your mix to a stereo audio file.

🕐 Duration

This exercise should take approximately 20 minutes to complete.

⊕ Goals/Targets

- Submix multiple tracks to an output bus channel
- Duplicate an Instrument track to create a variation on an existing MIDI performance
- Use the Tool Window to transpose notes to a different octave
- Export the final mix as a stereo file

Exercise Media

This exercise uses media files taken from the song, "Lights," provided courtesy of Bay Area band Fotograf.

Written by: Zack Vieira and Eric Kuehnl; Performed by: Fotograf

The media provided for this course may be used for educational purposes only. No rights are granted to use the media for any other personal, commercial, or non-commercial purposes.

Getting Started

To get started, you will open your completed project from Exercise 9. This will serve as the starting point for this exercise.

Open your existing Lights project:

1. Select **FILE > OPEN** or press **COMMAND+O** (Mac) or **CTRL+O** (Windows). The Browser will open.

2. Use the Browser to navigate to the folder containing your Lights-xxx project.

3. Select your Lights-xxx project from the list of available files.

4. Click **OPEN** at the bottom of the Browser. The project will open as it was when last saved. If needed, reopen the Main Mixer view (**WINDOW > VIEW MAIN MIXER**).

Creating a Drum Submix

You'll begin by creating a submix of all the drum tracks by routing them to a single output bus channel. You will then be able to use the output bus channel to set the volume for the drums as a whole without having to adjust the faders on all of the individual tracks.

Route the drums to an output bus submix channel:

1. Hold **COMMAND** (Mac) or **CTRL** (Windows) and click on each of your drum track channels (01 Kick, 02 Snare, 03 Hi Hat, and 04 Claps) to select them all.

2. Press **COMMAND+G** (Mac) or **CTRL+G** (Windows). Reason will create a new output bus Mix Channel device in the rack and a corresponding channel in the mixer. All of your selected drum tracks will be routed to the new channel.

3. The new output bus channel will have a default name, such as Bus 1. Double-click the name in the header of the output bus channel to rename the channel. Name the output bus channel Drum Submix. Your mixer channels should look similar to Figure 10.30.

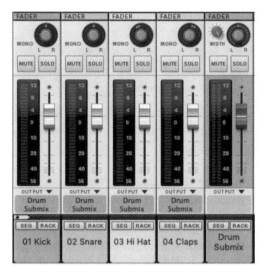

Figure 10.30 The drum tracks submixed into the Drum Submix output bus channel

You can now take a moment to adjust the level of the drums relative to other tracks in the mix using the Volume fader on the **Drum Submix** channel. Try moving the **Drum Submix** up or down a few dB to see how it changes the mix.

Adding a Second Strings Track

The **Strings** track sounds pretty good, but you may find that it sounds a little thin. In this section, you will create some more depth for the strings by adding a second track and transposing that track down an octave.

Duplicate the Strings track:

1. Click on the header of the **Strings** mixer channel to select it and verify that it is the only channel selected.

2. Do one of the following to duplicate the channel as well as the associated instrument track and instrument device:

    *   Select **EDIT > DUPLICATE CHANNELS AND TRACKS**.

    *   Right-click on the channel header and choose **DUPLICATE CHANNELS AND TRACKS**.

    *   Press **COMMAND+D** (Mac) or **CTRL+D** (Windows).

    The new strings mixer channel, instrument track, and instrument device will all be named **Strings copy**.

3.   Rename the new items by doing one of the following:

   • Using the Racks view, locate the NN-XT instrument device named Strings copy. Double-click on the device name to open a text box and change the name to Strings Low. Reason will automatically update the name of the sequencer track and the mixer channel.

   • Using the Sequencer view, locate the instrument track named Strings copy. Double-click on the track name to open a text box and change the name to Strings Low. Reason will automatically update the name of the instrument device and the mixer channel.

(i)   **If you rename the Strings copy mixer channel in the main mixer or the Mix Channel device in the rack, the name of the sequencer track and instrument device will not be automatically updated.**

4.   Using the process described in Step 3, rename the original Strings track, instrument, and mixer channel to Strings Hi.

5.   In the Main Mixer view, pan the Strings Low channel to the right a bit by setting the pan control to around 52. This will complement the panning you did in Exercise 7 on the Strings Hi track.

In the next series of steps, you'll transpose the duplicated notes to double the strings an octave lower.

Transpose the Strings Low track using the Tool Window:

1.   If the Sequencer view is not visible, press **F7** or choose **WINDOW > VIEW SEQUENCER** to display it.

2.   Using the Selection Tool, click on the clip at Bar 76 on the Strings Low note lane to select it.

3.   Do one of the following to display the Tool Window if it is not already visible:

   • Select **WINDOW > SHOW TOOL WINDOW**.

   • Press **F8**.

4.   If the Tool Window is not currently showing the Sequencer Tools tab, click on the Sequencer Tools button (wrench and screwdriver icon) in the top left of the window.

5.   Click the Fold/Unfold button next to the word Pitch to display the function, as needed.

6.   In the Pitch section, select the **TRANSPOSE (SEMITONES)** radio button, if not already active.

7.   Enter **-12** into the transpose field. This will transpose the MIDI part on the Strings Low track down by one octave.

8.   Click the **APPLY** button in the Pitch section to transpose the notes in the selected clip.

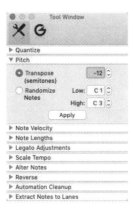

Figure 10.31 The Tool Window with the suggested Pitch transposition settings

Because you now have two strings tracks playing at the same time, the level of the strings in the mix may be a bit too loud. Try pulling the faders for both tracks down slightly, or generally adjust them until you find a balance that works in the mix.

 If you select multiple channels in the main mixer, the faders for all of the selected channels will move together when you move any one of them. You could also try submixing the strings tracks and adjusting the balance with the output bus fader.

Exporting an Audio Mix

In this section of the exercise, you will export the final audio mix of your project. Exporting a mix is a necessary step whenever you want to burn a song to CD, post it to a music-sharing site like SoundCloud, or sell it on services like iTunes or Band Camp. For this project, you'll create a 16-bit, 44.1 kHz WAV file. Selecting these parameters will ensure that the resulting file is compatible with all of the common distribution options.

Export your final mix:

1. Move the Song End Marker in the Ruler, placing it just past the end of the final fadeout that you created in Exercise 9, around Bar 103.

2. Select FILE > EXPORT SONG AS AUDIO FILE. The Export Song as Audio File dialog box will display.

Figure 10.32 The Export Song as Audio File dialog box (Mac)

Figure 10.33 The Export Song as Audio File dialog box (Windows)

3.   Specify the filename for your mix using the **SAVE AS** field (Mac) or **FILE NAME** field (Windows).

4.   If desired, specify a different location for your exported mix using the folder navigation controls in the dialog box.

5.   Using the **FORMAT** pop-up menu (Mac) or **SAVE AS TYPE** pop-up menu (Windows), choose **WAVE FILE** as the output format.

6.   Click the **SAVE** button to continue to the Audio Export Settings dialog box.

Figure 10.34 The Audio Export Settings dialog box

7. In the Audio Export Settings dialog box, select the following settings:

- Sample Rate: **44,100 Hz**

- Bit Depth: **16**

- Dither: **ENABLED**

8. When ready, click the **EXPORT** button to begin exporting your mix. You will see a progress bar in the Export Audio window as the export completes.

Finishing Up

Congratulations! You've completed all required audio production work for a Reason project, including:

- Importing audio and MIDI files

- Editing clips and creating virtual instruments

- Mixing and balancing tracks

- Adding EQ, dynamics, and effects processing

- Creating automation

- Submixing tracks

- Applying note changes with the Tool Window

- Exporting the final mix as a stereo file

To wrap up, you'll need to locate the file you exported and verify the results.

Locate your exported stereo mix file:

1. Switch from Reason to the Mac Finder or Windows File Explorer.

2. Navigate to the folder location you selected for your export.

3.   Select your mix file and do one of the following to listen to the results of your exported mix:

- On a Mac-based system, press the **SPACEBAR** to begin playback using the QuickLook feature.

- On a Windows-based system, right-click on the file and select **PLAY WITH WINDOWS MEDIA PLAYER**.

4.   Listen through the entire mix to verify that the results are as you expected/intended.

 Be sure to listen to mixed files that you export before sharing them with others to check for any unintended condition that may have affected the export.

5.   If you hear any problems with the file (such as missing parts, wrong file duration, etc.), return to Reason to correct the issue. Look for soloed or muted tracks, an incorrect Song End Marker location, and so forth. Once corrected, play back the project to verify that it sounds correct and then repeat the export process.

That completes this exercise.

Index

About the Author

This book was written by Zac Changnon and published under the direction of NextPoint Training, Inc. The course is designed to prepare students for certification in basic audio production using Reason software under the NextPoint Training Digital Media Production program.

Zachary Changnon is a producer and longtime Reason user, starting on version 2 of the software. He is currently working as a freelance author and composer based in the Los Angeles area.

Previously, Changnon was a software developer for thirteen years with Hyland Software in Westlake, Ohio. He left that position to study audio technology and production in depth at Pyramind in San Francisco. He holds a bachelor's degree from the University of Toledo, Ohio, a Complete Producer program certificate from Pyramind, and a Pro Tools|Music Expert certification from Avid.